Damián Ezequiel Alonso

Wireless Data Transmission for the Battery Management System of Electric and Hybrid Vehicles

Forschungsberichte aus der Industriellen Informationstechnik
Band 15

Institut für Industrielle Informationstechnik
Karlsruher Institut für Technologie
Hrsg. Prof. Dr.-Ing. Fernando Puente León
Prof. Dr.-Ing. habil. Klaus Dostert

Eine Übersicht aller bisher in dieser Schriftenreihe erschienenen Bände finden Sie am Ende des Buchs.

Wireless Data Transmission for the Battery Management System of Electric and Hybrid Vehicles

by

Damián Ezequiel Alonso

KIT Scientific Publishing

Karlsruher Institut für Technologie
Institut für Industrielle Informationstechnik

Wireless Data Transmission for the Battery Management System of Electric and Hybrid Vehicles

Zur Erlangung des akademischen Grades eines Doktor-Ingenieurs von der Fakultät für Elektrotechnik und Informationstechnik des Karlsruher Instituts für Technologie (KIT) genehmigte Dissertation

von Ing. Damián Ezequiel Alonso aus Buenos Aires, Argentinien

Tag der mündlichen Prüfung: 08. August 2016
Referenten: Prof. Dr.-Ing. habil. Klaus Dostert
Prof. Dr.-Ing. Thomas Zwick

Impressum

KIT Scientific Publishing

Karlsruher Institut für Technologie (KIT)
KIT Scientific Publishing
Straße am Forum 2
D-76131 Karlsruhe

www.ksp.kit.edu

Print on Demand 2017 – Gedruckt auf FSC-zertifiziertem Papier

ISSN 2190-6629
ISBN 978-3-7315-0670-6
DOI 10.5445/KSP/1000069560

Abstract

Battery management systems (BMSs) are designed to monitor and control the traction battery in hybrid and electric vehicles. A communication network is established within the battery, in order to collect the information measured by sensors distributed among the battery cells. State-of-the-art BMSs employ wired bus systems, such as the CAN bus. This leads to the installation of large wiring harnesses, which increases costs and manufacturing complexity. This thesis introduces a novel wireless approach for the data transmission within the BMS, which consists of installing an antenna in each communication device of the BMS.

Two antenna types operating in two different frequency ranges have been chosen as candidates. The investigation of the wireless in-battery channel is first realized by means of measurements in self-designed channel emulators, and then expanded by means of software-based electromagnetic simulations. The influence of different system parameters on the channel is analyzed, such as physical dimensions, and number and position of the antennas.

A simple architecture for the communication modems is proposed. The entire communication system is first simulated in software and then implemented by means of rapid prototyping platforms. The performance is measured in the channel emulator and also in a real automotive battery.

The results of channel measurements and simulations, as well as the communication system performance, confirm the technical feasibility of the wireless data transmission for the BMS. The necessary transmission power is very low in every considered case. Working at low frequencies is a good alternative for a reduced number of antennas, due to the lower dependence of the studied system parameters on the channel characteristics. However, operating at higher frequencies is a better approach for larger numbers of antennas: The transmission power is lower and the suggested type of antenna is smaller and cheaper.

Zusammenfassung

Batteriemanagementsysteme (BMS) werden für die Überwachung und Steuerung der Traktionsbatterien in Hybrid- und Elektrofahrzeugen entworfen. Ein Kommunikationsnetzwerk wird in die Batterie eingesetzt, um die Informationen zu sammeln, die von Sensoren zwischen den Batteriezellen gemessen werden. Aktuelle BMS verwenden verdrahtete Bussysteme wie CAN-Bus. Dies führt zur Installation von großen Kabelbäumen, was die Kosten und den Herstellungsaufwand erhöht. In dieser Arbeit wird ein neuartiger drahtloser Ansatz für die Datenübertragung innerhalb des BMS dargestellt, der die Installation einer Antenne bei jedem Kommunikationsgerät des BMS vorsieht.

Zwei Antennenarten für zwei verschiedene Frequenzbereiche wurden als Kandidaten ausgewählt. Die Untersuchung des drahtlosen in-Batterie-Kanals wird zuerst durch Messungen in selbstentworfenen Kanalemulatoren durchgeführt und danach mittels softwarebasierten elektromagnetischen Simulationen erweitert. Der Einfluss verschiedener Systemparameter auf den Kanal wird untersucht , z. B. die physikalischen Dimensionen und die Anzahl und Position der Antennen.

Eine einfache Architektur wird für die Modems des Kommunikationssystems vorgeschlagen. Das gesamte Kommunikationssystem wird zuerst in einer Software simuliert und anschließend mittels Rapid-Prototyping-Plattformen implementiert. Die Leistung wird im Kanalemulator sowie auch in einer realen Fahrzeugbatterie gemessen.

Die Ergebnisse der Kanalmessungen und Simulationen sowie die Leistung des Systems bestätigen die technische Machbarkeit der drahtlosen Datenübertragung für das BMS. Die notwendige Sendeleistung ist sehr gering in jedem betrachteten Fall. Aufgrund der geringen Abhängigkeit der Kanaleigenschaften von den untersuchten Systemparametern ist das Arbeiten bei niedrigen Frequenzen eine gute Alternative für eine reduzierte Anzahl von Antennen. Aufgrund der wesentlich niedrigeren erforderlichen Sendeleistung sowie billigeren und kleinern Antennen sind höhere Frequenzen für große Antennenzahlen ein besserer Ansatz.

Acknowledgements

"We are like dwarves perched on the shoulders of giants"
(Bernard of Chartres)

First of all, I would like to sincerely thank my supervisor Prof. Dr.-Ing. habil. Klaus Dostert for the opportunity to carry out my PhD thesis at the Karslruhe Institute of Technology (KIT), as well as his continuous and valuable support to my work. I would also like to acknowledge Prof. Dr.-Ing. Thomas Zwick for the co-supervision of my work and his relevant advice to improve my thesis.

I am deeply grateful to all my colleagues and the personal staff at the Institute of Industrial Information (IIIT) for the friendly working atmosphere, as well as the good advice and cooperation. I would really like to particularly thank my colleague Oliver Opalko for his valuable collaboration within the IntLiIon project, his dedication to revise my thesis, and for being an awesome office mate.

I have to pleasure the cooperation of the partners from Bosch, ProDesing and the Hochschule Hannover along the IntLiIon project. I specially owe gratitude to Thomas Buck and Wenqing Liu for giving me access to the Bosch Battery Labs, where I could demonstrate the conclusions of my investigation in a real automotive battery.

I am grateful to my students, who contributed to my research with their theses or working as HiWis. It has been a pleasure to work together.

I want also to thank all the wonderful friends that I met during these years in Karlsruhe and made me really enjoy my time in this city.

Finally but most important, I want to express my infinite gratitude to my parents and my sister. Every step forward in my life, does not matter how big or small it is, is thank to the values learned at home, as well as their continuous support and love despite the long distance.

Contents

Nomenclature

Common abbreviations

Abbreviation	Description
amp.	amplifier
cod.	encoded
cont.	continuos
demod.	demodulator
e.g.	exempli gratia
etc.	et cetera
i.e.	id est
meas.	measurement
mod.	modulator
recov.	recovery
sync.	synchronization
unc.	uncoded
ACF	Auto-correlation function
ADC	Analog to digital converter
AGC	Automatic gain control
ASK	Amplitude-shift key
AWGN	Additive white Gaussian noise
BB	Baseband
BCU	Battery control unit
BER	Bit error rate
BMS	Battery management system
BPSK	Binary phase-shift keying
CAD	Computer-aided design
CAN	Control area network
CDMA	Code division multiple access
CIR	Channel impulse response
CSC	Cell supervision circuit
CTF	Channel transfer function
DAC	Digital to analog converter
DQPSK	Differential quadrature phase-shift keying
DSP	Digital signal processor
EM	Electromagnetic

Abbreviation	Description
EOB	End of burst
FDMA	Frequency-division multiplexing
FEC	Forward error correction
FFT	Fast Fourier transform
FGPA	Field programmable gate array
FIR	Finite impulse response
FSK	Frequency-shift key
GPL	General public license
GPS	Global positioning system
I/Q	Inphase/quadrature
IDFT	Inverse discrete Fourier transform
IEEE	Institute of Electrical and Electronics Engineers
IFFT	Inverse fast Fourier transform
IIIT	Institute of Industrial Information Technology
IntLilon	Intelligent data bus concepts for lithium-ion batteries in electric and hybrid vehicles
ISI	Inter-symbol interference
ISM	Industrial, scientific and medical
KIT	Karlsruhe Institute of Technology
LFSR	Linear feedback shift registers
MAC	Media access control
Mbps	Megabits per second
MDO	Mixed domain oscilloscope
MLS	Maximum length sequences
MSC	Module supervision circuit
MSps	Mega-samples per second
NTP	Network time protocol
OFDM	Orthogonal frequency division multiplexing
PCB	Printed circuit boards
PDP	Power delay profile
PIFA	Planar inverted-F antenna
PLC	Power line communication
PN	Pseudo noise
PSK	Phase-shift keying
QPSK	Quadrature phase-shift keying
RMS	Root mean square
RRC	Root-raised-cosine
Rx	Receiver
SDMA	Space division multiple access

Abbreviation	Description
SDR	Software-defined radio
SF	Similarity factor
SMA	Subminiature version A
SNR	Signal-to-noise ratio
SOB	Start of burst
SoC	State-of-charge
SoH	State-of-health
sps	Samples per symbol
TDMA	Time division multiple access
TDM	Time-division multiplexing
Tx	Transmiter
UHD	USRP hardware driver
USRP	Universal software radio peripheral
VHF	Very High Frequency
VNA	Vector network analyzer
WLAN	Wireless local area networks
XCF	Cross-correlation function

Letters

Latin Letters

Symbol	Description
B	Bandwidth
B_{coh}	Coherence bandwidth
B_{sym}	Symbol bandwidth
c	Speed of light
d	Distance between antennas
E_X	Energy of the function X
E_{b}/N_0	Energy per bit to noise power spectral density ratio
$(E_{\text{b}}/N_0)_{(\text{ADC})}$	Energy per bit to noise power spectral density ratio calculated with the measurements of the ADC
$(E_{\text{b}}/N_0)_{(\text{Rx})}$	Energy per bit at the receiver to noise power spectral density ratio
$(E_{\text{b}}/N_0)_{(\text{Tx})}$	Energy per bit at the transmitter to noise power spectral density ratio
f	Frequency

Symbol	Description
f_c	Center frequency
$f_{\mathrm{min}(l>0)}$	Minimum resonant frequency with resonant mode $l > 0$
f_{mnl}	Resonant frequency of the resonant mode (m, n, l)
f_s	Sampling frequency
h	Channel impulse response in time domain
H	Height of the battery emulator
H_ch	Coefficients of the CTF in frequency domain
H_RxF	Coefficients of the receive filter in frequency domain
H_TxF	Coefficients of the transmit filter in frequency domain
k	Boltzmann's constant
L	Length of the battery emulator
L_a	Length of the antenna
L_frame	Frame length
L_payload	Length of the frame payload
M	Modulation order
N_cm	Number of cells in a module
N_m	Number of modules in a battery
N_0	Noise power spectral density
N_s	Number of slaves
P	Recieved power
P_Tx	Transmission power
R_b	Gross bit rate
R_bN	Net bit rate
R_s	Symbol rate
S_{mn}	Scattering parameter (m, n) from a Scattering matrix
T	Temperature
T_frame	Time to transfer a frame
T_sym	Symbol duration
W	Width of the battery emulator
x	X-coordinate
y	Y-coordinate
z	Z-coordinate
Z_a	Height of the antenna in the battery emulator

Greek Letters

Symbol	Description
β	Roll-off factor
δ	Electromagnetic penetration depth
Δf	Frequency offset
ϵ_r	Relative permittivity
λ	Wavelength
μ_r	Relative magnetic permeability
ω	Angular frequency
σ	Electric conductivity
$\star$	Cross-correlation function operation
$\bar{\tau}$	Mean delay spread
σ_τ	RMS delay spread

1 Introduction

1.1 Motivation

Lithium-ion batteries are currently the most employed technology for the available automotive batteries (also named traction batteries) on the market, due to their high energy density, long service life and environmental friendliness [39, 48]. The growing market of electric and hybrid vehicles has promoted the research on new technologies for the optimization of automotive batteries. The aims are always related to the reduction of cost, weight, size and manufacturing complexity, as well as the increase of reliability and performance [27].

Automotive lithium-ion batteries operate without facing problems related to their durability and safety only under certain restrictive conditions [48], regarding aspects such as operating temperature, operating voltage, charge balance, etc. Therefore, a battery management system (BMS) is required to guarantee and supervise these conditions. The BMS is in charge of monitoring and controlling functions within traction batteries. Some essential functions are the measurement, control and optimization of the state-of-charge (SoC) and state-of-health (SoH), the charge balance between cells, and the detection of failures (e.g. cell short circuits, temperatures too high, etc.) [100].

The performance of automotive batteries can be boosted in different ways. Improving the technology employed in the battery cells, as well as the battery manufacturing process, has been a hot topic for a long time and it has been largely investigated [19]. Another alternative is enhancing the management system [8]. Some aspects to be considered in this sense are the available bit rate within the battery sensor network, as well as the reduction of the extra weight and costs, generated by the deployment of the BMS.

A typical automotive battery is represented in Fig. 1.1, which presents the most important components of the battery that are explained in this subsection. The key components of an automotive battery are the cells. They generate electrical energy by means of a chemical reaction. The number of cells in commercial batteries can vary from quantities smaller than one hundred [84] to several thousands [24, 48, 54].

A large number of sensors are distributed over the battery in order to perform the previously described monitoring tasks. Each battery cell is equipped with a cell supervision circuit (CSC), which includes sensors for different physical variables, such as temperature and voltage. The BMS also has a central device, commonly named battery control unit (BCU), responsible for collecting the information sensed in the whole battery. Then, the BMS processes this information and executes the required control tasks.

Since the voltage and capacity of single battery cells are usually small, a certain number of cells is connected in series and packed into a module [48]. These modules are connected in series to reach a certain battery voltage. Sometimes also even in parallel to deliver a higher current. Logically, the larger the number of cells, the higher the maximum power that the battery reaches. But this also increases the cost, size and weight of the battery. Therefore, a good compromise must always be found.

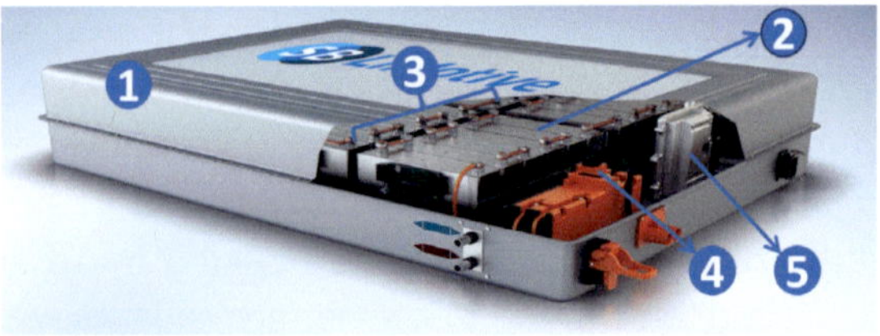

Figure 1.1: Sketch of an automotive lithium-ion battery (source: SB Limotive)

1.2 In-Battery Communication

A communication network is deployed in every automotive battery as part of the BMS. In this way, the BCU can communicate with every CSC to read its information and also send them commands, if necessary. The data transmission within the BMS will be referred as *in-battery communication* in the remaining work. There are two possible ways of communication between the BCU and the CSCs: cell-wise or module-wise. In the first case, the BCU can communicate directly with each individual CSC. In the case of module-wise communication, the BCU does not communicate directly with the CSCs, but with a module supervision circuit (MSC), which is included in every battery module [24, 48]. Each MSC previously collects and stores the information from the CSCs of its cells.

In current commercial BMSs, the data transmission is realized by means of standard wired bus systems. Control area network (CAN) bus is the most used method [24, 99, 100]. Such bus systems employ a master-slave architecture. Thereby, the BCU acts as master, and the slaves are either the MSCs, in the case of module-wise communication, or the CSCs, in the case of cell-wise communication.

1.2.1 State of the Art and Problem Statement

The use of wired buses requires the installation of a very large wiring harness along the battery pack, in order to interconnect the large number of communication nodes [99, 100]. This leads to an increase of costs and weight. The battery assembly process is also more complex, due to the installation of the wiring and the galvanic isolation between battery cells and cables. Besides, the communication wiring has a negative effect on safety because it is susceptible to mechanical failures, due to fatigue or vibration [81]. Finally, the CAN bus limits the maximum bit rate to 1 Mbps, according to the norm ISO 11898-2 [34]. This could turn into a bottle neck for the development of future generations of battery sensor systems.

With the aim of keeping the size of the wire harness as small as possible, a module-wise communication is normally preferred for the BMS. The negative consequence is that sometimes the entire battery module must be changed even if only one single battery cell is damaged.

1.3 The Wireless Approach

1.3.1 Framework of the Project

The IntLilon ("Intelligent data bus concepts for lithium-ion batteries in electric and hybrid vehicles") project aims to develop novel and smart data transmission techniques for the BMS [77]. The focus of this research is on innovative communication systems being capable of efficiently controlling and monitoring the battery, but avoiding the necessity of the problematic additional wiring for communication purposes. Furthermore, the feasibility of a cell-wise monitoring is also investigated. This could allow the replacement of single cells instead of complete modules in case of malfunction, saving resources and bringing down costs.

The task assigned to the Institute of Industrial Information Technologies (IIIT) of the Karlsruhe Institute of Technology (KIT) is the investigation of different alternatives for the physical layer of the in-battery communication network. Two different approaches have been studied under the framework of the IntLilon project: power line communication (PLC) and wireless communication. This thesis is focused on the latter.

1.3.2 Goals and Challenges

The novel approach investigated in this work proposes installing an antenna in each communication node of the battery system to facilitate the wireless communication between them. That means, an antenna must be placed in the master (BCU) and in every slave, either CSCs or MSCs. In this way, the dedicated wiring for communication purposes within the battery is completely avoided. The electromagnetic waves will propagate along the free space inside the traction battery, mainly between the upper side of the cell cans and the battery housing. A representation of the wireless cell-wise communication is presented in Fig. 1.2(a) and the wireless module-wise communcation in Fig. 1.2(b).

In this work, the wireless channel inside an automotive battery is investigated. The focus is on the channel transfer function (CTF) between the master and slave antennas, and how these channels are affected by different parameters, such as the battery size, the operating frequency range, and the antenna positions. This analysis is carried out by means

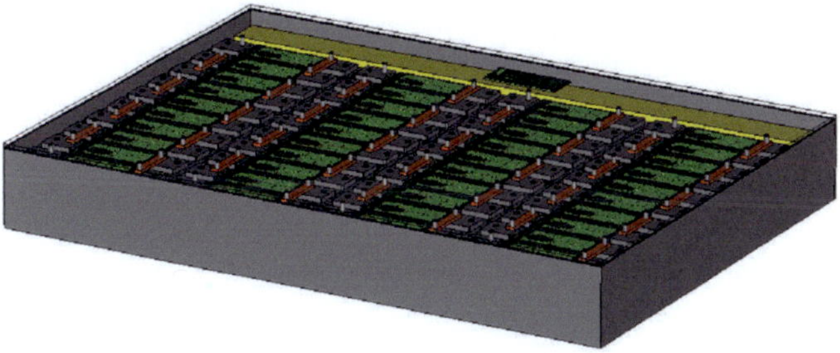

(a) Cell-wise communication.

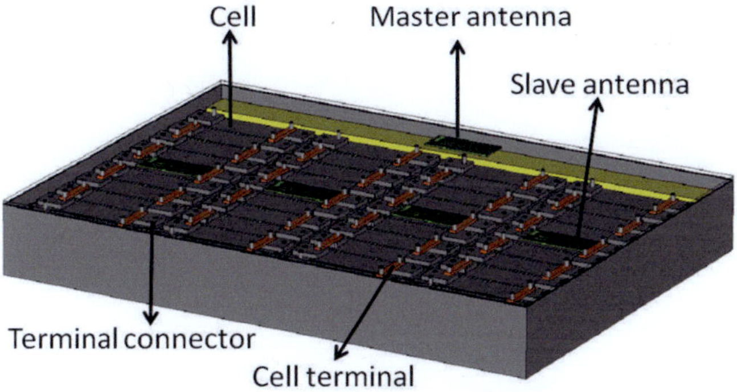

(b) Module-wise communication.

Figure 1.2: Illustration of cell-wise and module-wise communication architectures

of measurements in a channel emulator and also by software-based electromagnetic (EM) simulations. Then, based on these results, different transmission methods are investigated, in order to design a communication system, according to the channel and project requirements. The complexity of this system must be kept as low as possible in order to reduce the costs. The latter is particularly important in the case of cell-wise communication, due to the large amount of modems that must be deployed. Finally, a laboratory demonstrator of the proposed approach is implemented to prove the feasibility of the proposal.

One important challenge of this approach is that, as it is explained in Section 1.5, no published work related to the wireless channel inside automotive batteries was available before the beginning of this doctoral thesis. An environment as the interior of an automotive battery, where the height of the available space is only some centimeters, and most of the surrounding objects are reflective (because they are made of metal), is also very uncommon for a wireless data transmission based on antennas. Additionally, the size of the battery and distance between the antennas are in a similar order of magnitude as the wavelength of the frequency ranges under investigation. This leads to a resonant behavior of the channel.

1.3.3 Scope of the Work

The following constrains have been assumed under the framework of the IntLiIon project for the case of the wireless approach:

- the battery has a metallic housing
- the battery contains between 50 and 200 cells
- each module has between 8 and 14 cells
- the cell cans are made of metal
- the BCU and the battery cells are contained in an unique prismatic compartment and distributed along only one plane

Regarding the last point: If any slave would be placed in a different metallic compartment as the master, the wave propagation between them would be practically impossible. The same would happen if the cells were distributed in several parallel planes. In all these cases, however, one master could be installed above each battery cell plane in each compartment. Then, the masters could communicate between them by means of one of the currently employed wired bus systems. But even in such cases, the communication inside each compartment level would be similar to the one explained in this work, so all the research carried out would be still valid.

1.4 Project Specifications and Requirements

Under the frame of the IntLiIon project, the following general specifications have been agreed between the project members:

- Minimum bit rate: 2 Mbps (twice the maximum of the CAN bus)
- Maximum bit error rate (BER) of 10^{-3} without employing channel coding and 10^{-6} after channel coding
- The physical variables to be measured by the system at each cell are the following:
  - Voltage: Resolution of 14 bits, maximum latency of 20 ms between subsequent readings (a maximum latency of 10 ms is desirable), measurement range: from 2.7 V to 4.2 V
  - Temperature: Resolution of 11 bits, maximum latency of 100 ms between subsequent readings, measurement range: from −40 °C to 50 °C
- A master-slave architecture is employed: The master is the BCU, and the slaves are either the CSCs (in case of cell-wise communication) or the MSCs (in case of module-wise communication)
- The communication between master and slave must be half-duplex and direct, i.e. without repeaters or multi-hopping techniques; communication between slaves is not required
- A time division multiple access (TDMA) protocol with master-slave hierarchy must be employed for the media access control (MAC) layer

Additional specifications have also been agreed. They are, however, not directly connected with the scope of this thesis, as for example the functional safety. Therefore, they are not treated in this thesis. Further information can be found in [85].

In the particular case of the wireless approach, some extra requirements must be taken into account for the design of the antennas and their associated modems (the communication transceivers):

- Low cost: As already mentioned, one of the main targets of IntLiIon is to reduce costs. In particular, the cost of the antennas and the modems is a critical factor, due to the big amount of them that would be deployed in the battery
- Small size: The free space available among cells and battery housing is normally very small. This must be, however, large enough to allow the placement of antennas and modems
- Suitability: The same antenna and modem must be employed at every slave or master position along a battery. Ideally, the same antenna and modem should also properly operate in different batteries

1.5 Related Work

Some studies of wireless in-battery communication were already presented. Different authors investigated the application of the standard IEEE 802.15.4 in [55, 97]. However, this standard allows a maximum bit rate of only 250 kbps, even less than the CAN bus. Under the framework of the project BATSEN, the authors dealt with the development of sensors [80] and also antennas [79] for a wireless battery management system. Additionally, Lee et al. [40] designed an integrated circuit for the same purpose and they defined a communication protocol (WiBaAN). However, to the best of the author's knowledge, there is no author who has characterized or modeled the wireless in-battery channel before this thesis.

As already mentioned, other approaches based on power line communication were also studied under the framework of the IntLiIon project. A parallel coupling technique was employed by Ouannes et al. [65–67]: The connection between each communication modem an the cells or the BCU is parallel to the cell or BCU terminals. Opalko et al. [108–111] studied a serial coupling technique, by means of Rogowski coils. In both cases, the power line channel was modeled, as well as the performance of different communication system approaches under these channels by means of simulations.

1.6 Structure of the Thesis

The thesis is organized as follows: Chapter 2 explains how the emulators of the wireless in-battery channel were conceived and developed. In Chapter 3, the first channel measurements are carried out in the channel emulators and the antenna candidates for this work are selected. Chapter 4 introduces the models for the software-based electromagnetic simulations. in Chapter 5, the wireless in-battery channel is analyzed and characterized. Chapter 6 proposes a design of the communication modem and evaluates the performance results in the investigated channels by means of simulations. Chapter 7 presents the modem prototypes developed by means of software defined radio platforms and performance of these prototypes in the channel emulator. Chapter 8 shows the results of the measurements carried out in an cold automotive battery. Finally, Chapter 9 concludes this doctoral thesis and raises discussion for future work.

2 Measurement Testbed of the Wireless In-Battery Channel

2.1 Wireless In-Battery Channel Emulator

For the deployment of any communication system, the knowledge of the communication channel is a key factor. As already explained in Section 1.5, practically no research results were available about the wireless in-battery channel before this doctoral thesis. The first investigation of the channel was carried out by means of measurements. However, these measurements were not performed in a real "hot" [1] automotive battery, which can produce explosions if they are not handled properly [2], due to the very high energy density storage of the cells. Therefore, a measurement in a hot battery could be very complicated and risky. Another alternative would be to conduct the measurements in a "cold" battery. This should not significantly modify the results that would be obtained in a hot battery, since no component of the power train produces significant additional noise in the investigated frequency range for the wireless approach [16].

Nonetheless, measuring in a real automotive battery would have been impractical, since the preparation of every new measurement campaign would require a long time. Additionally, our group did not have a real automotive battery at the beginning of the project. Therefore, it was necessary to design an emulator of the wireless channel within automotive batteries. The emulator is denoted as "channel emulator"' or "battery emulator" [2] in the remaining work. The idea behind this emulator has been to emulate the effect of the battery components that would have the highest impact on the communication channel between antennas in the wireless approach.

1 An automotive battery is named "hot" when the battery cells include the conducting fluid. In contrast, it is called "cold" when the cells do not have the conductive fluid and, therefore, they can neither be charged nor discharged.

2 "Battery emulator" has been the employed denotation up to now in all the publications related to this wireless approach [102, 104–107].

There is a large amount of different architectures on the market. Since the battery geometry is adapted to the characteristics of every particular vehicle, there are no standards for the geometry or sizes of automotive batteries. Some batteries have a compact design, containing all the modules in a single compartment (the battery housing). Other battery systems are divided into multiple sub-packs, distributed over the vehicle, exploiting the available space more efficiently [100]. As remarked in Section 1.3.3, in this work only the first case is considered, with every cell on the same plane as represented in Fig. 1.1. In this way, the battery cells are physically distributed in a two-dimensional matrix. Each cell is placed very close to the adjacent cells and/or to the lateral housing walls. Therefore, the upper sides of all the cells are treated as a virtual metallic plane, since the free space between cells and/or housing lateral walls is much smaller than the wavelengths corresponding to the frequency range under study (the latter is described in Section 3.1). Under these considerations, as the antennas must be placed above the cells or the modules, the electromagnetic waves propagate practically only along the free space between the virtual metallic plane and the metallic battery housing above it. This particular environment can be described as a rectangular metallic prism, which has the same length and width as the battery housing, but its height is the distance between the cells' upper sides and the top of the housing.

Two channel emulators were built, using the physical dimensions of commercial automotive batteries as reference. The dimensions of the channel emulators[3] are listed in Table 2.1. The emulators have been already presented in [107]. Both are made of stainless steel, with a thickness of 1.5 mm. Fig 2.1 presents a photo of the small battery emulator without the top cover and with five antennas inside. All the measurements are, however, always performed with the emulator closed by its cover.

[3] In the works [102, 104, 107], it was mentioned that the small battery emulator dimensions were 800 mm in length and 500 mm in width. These were actually the dimensions of the reference battery, but the real dimensions of the manufactured emulator are documented in the current work.

Table 2.1: Dimensions of the manufactured channel emulators

Dimension	Small Emulator	Big Emulator
Length (L) [mm]	785	1700
Width (W) [mm]	475	1000
Height (H) [mm]	52	52

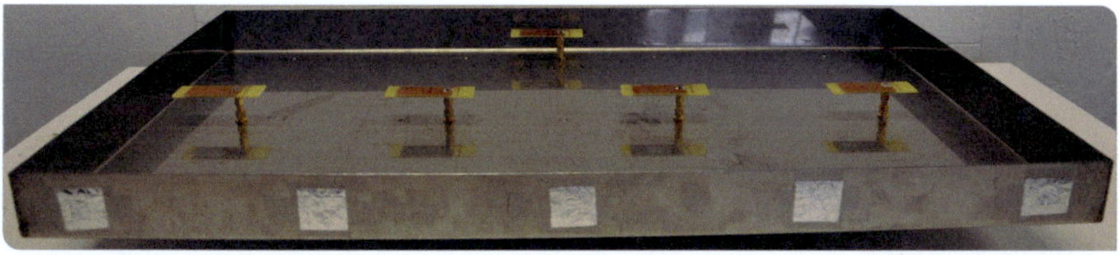

Figure 2.1: Photo of the small battery emulator without top cover, with five antennas inside

2.2 Comparison With Ideal Cavity Resonators

The channel emulators have many similar characteristics to ideal rectangular cavity resonators. The type of resonators has already been extensively investigated [71, 90]. The channel emulators can be described, as well as the mentioned resonators, as a piece of a rectangular waveguide short-circuited at both ends.

A perfect electric conductor material is assumed for the walls of ideal resonators in most literature. In the channel emulator, the walls are made of a material with very high conductivity. In order to analyze the effectiveness of the emulator walls to block electromagnetic waves within the frequency range of interest, the electromagnetic penetration depth (δ) of the emulator walls must be calculated. In the case of conducting materials, δ can be obtained as [90]

$$\delta = \sqrt{\frac{2}{\omega \cdot \mu \cdot \sigma}}, \tag{2.1}$$

where $\omega = 2 \cdot \pi \cdot f$ (f is the frequency); μ is the magnetic permeability and σ is the electric conductivity of the material. In the case of the channel emulator walls $\mu \approx \mu_o = 4\pi \times 10^{-7}$ H/m and $\sigma = 1.45 \times 10^6$ S/m. The highest value of δ appears for the lowest frequency. As later explained in Section 3.1, the lowest frequency value investigated in this work is 100 MHz, which results in $\delta = 41.80\,\mu\text{m}$. Then, since δ is much smaller than the channel emulator walls thickness for the complete frequency range of interest, it can be concluded that the emulator walls block any electromagnetic radiation along the frequency range of interest. Consequently, the interior of the emulator can be considered as electromagnetically isolated, i.e. shielded from the outside.

An important difference between the emulated channel and ideal cavities are the excitation sources. The excitation sources are ideal in most literature, i.e. the sources only generate an electric or magnetic field inside the cavity, but they do not modify the wave propagation characteristics inside it. However, the size of the antennas cannot be neglected in comparison with the emulator dimensions in the wireless approach under study in this work. Hence, they modify the resonant modes inside the emulator. Nevertheless, many concepts from the studies on rectangular cavity resonators are helpful to explain several characteristics of the wireless channel inside the emulator.

2.3 Antenna Connection and Orientation

Several antennas have been placed and measured inside the channel emulators, as presented in Chapter 3. These antennas have been connected to different laboratory instruments, such as a vector network analyzer (VNA), or a spectrum analyzer. However, accurately measuring small antennas is normally not an easy task. The feed cables are coaxial cables with a single-ended connection, and they can strongly modify the measurement result [45]. The feed cables reflect and scatter the electromagnetic field produced by the antenna, modifying the antenna's radiation pattern. Additionally, the feed cable produces secondary radiation, due to the common-mode current that flows on its outer surface. Weng et al. [94] also reported that normally it is more difficult to reach a good agreement between measurements and simulations when the ground plane is electrically small[4]. This is the case in many of the antennas under study, especially the ones for the lowest frequencies of interest. Consequently,

4 Electrically small or large means that a particular physical dimension is considerably smaller or larger, respectively, in comparison with the wavelength under consideration.

the measurement results can be significantly affected by, for example, the position and length of the cables. This leads to inaccurate and also hardly reproducible results.

The described problem has been largely studied by other authors, and many alternatives have been presented [46, 49]. In the battery emulator employed for the wireless approach, the problem was solved by a particular connection method between the antennas and the emulator. All the measured antennas have a single-ended feeding point. The latter is connected to a Subminiature version A (SMA) adapter, which is screwed to the emulator floor, as shown in Fig. 2.2(a). The feeding coaxial cable is connected at the other end of the adapter, located at the opposite side of the floor, outside the emulator, as depicted in Fig. 2.2(b). In this way, the cable is shielded from the antenna by the metallic floor of the emulator [102] and a good electrical connection between the antenna ground plane and the emulator is guaranteed. At this side of the adapter, different elements can be connected to the antennas, such as laboratory instruments or communication platforms.

The antennas are always placed with their ground planes parallel to the floor of the emulator or to the upper side of the cells, in the case of a real battery. This is done in this way because in an automotive battery the height of the free space above the cells is very small, as already mentioned. This height could be even only around 2 cm, i.e. shorter than the length of every antenna under study in this work. The consequences of the antenna orientation are discussed in Section 3.6.

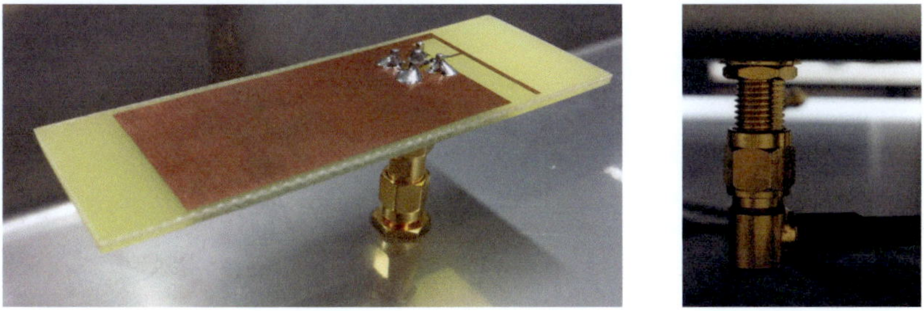

(a) Above **(b)** Under

Figure 2.2: Connection between the antenna and the battery emulator, above (a) and under (b) the emulator floor

2.4 Setup for CTF Measurements

The most important channel characteristic for the analysis of the wireless in-battery channel is the CTF between antennas placed at different positions inside the channel emulator, emulating the master and slave antennas. The laboratory instrument used to measure these CTFs was a VNA. This instrument measures the amplitude of the wave quantities (incident and reflected waves) between the ports of an electrical network. Based on the wave quantities, it calculates different network parameters, such as the scattering parameters (S-parameters) [32]. The VNA has been always calibrated to measure from the end of the feed cable, where the SMA adapter is connected.

Most of the components engaged in communication system front-ends for wireless communications are matched to 50 Ω. The same applies to the employed VNA. Hence, the S-parameter between two antennas represents the CTF between them. For example, S_{21} is the CTF from the antenna connected to port 1 to the antenna connected to port 2. As the measured network consists only of passive elements and the employed materials can be assumed to be isotropic, the network is reciprocal. That means that the S-parameter matrix is symmetrical, i.e. $S_{mn} = S_{nm}$ [71].

The employed VNA has four ports, so it can measure the S-parameters between a maximum of four antennas at the same time. However, a system with more antennas can be also easily measured: As only the channels between the master and each slave (but not between slaves) are of interest, and the channels under measurement are assumed to be time-invariant, all CTFs can be obtained by subsequent measurements, changing the slaves connected to the VNA. The unconnected slaves in each measurement must be terminated with 50 Ω, since in the proposed system the antennas would be always connected to a modem, whose input/output impedance would be also 50 Ω.

The described setup is employed in the next chapter for the CTF measurements between antennas. In Chapter 4 is modeled for electromagnetic simulations.

3 Channel Measurements and Selection of the Antennas

The type of antenna and the operating frequency range are two key aspects of the wireless approach for the BMS. They have a predominant effect on the CTF between the antennas installed in the communication nodes. However, carrying out a deep analysis of each antenna candidate for the in-battery communication would probably demand a time that would exceed the project time constrains. Therefore, a test method was developed and executed in order to perform a faster and simpler pre-selection of the antennas.

Several antenna types designed for different frequency ranges have been measured inside the channel emulator in different positions. Based on these results, the antennas with the best performance have been selected for the next steps of this work. This chapter describes first the different types of antennas under test and the frequency ranges that were considered. Then, the CTFs measured inside the channel emulator are presented and compared. The complications by measuring antennas at high frequencies in our emulator are explained. Finally, the selection of the antennas is discussed.

3.1 Investigated Frequency Range

The wireless in-battery communication is a novel approach still under investigation and, therefore, it has been never regulated. That means that there is no frequency band along the spectrum reserved for this type of communication [56]. Consequently, it was decided to first investigate a large frequency range, in order to make the analysis as extensive as possible.

The industrial, scientific and medical (ISM)frequency bands have been used as reference in order to determine the limits of the frequency range under investigation. The ISM bands are portions of the spectrum that are not regulated for telecommunications, but for general purposes [35]. They are listed in Table 3.1. The first four lowest ISM bands have been immediately discarded, because they would require antennas whose size would be inadmissible for the free space available inside a automotive battery. Only the ISM bands from the fourth one (from 433.05 MHz) have been considered. Nonetheless, it was decided to extend the bottom limit of the frequency range to be investigated in order to include the Very High Frequency (VHF) Band III, which in Europe ranges from 174 MHz to 230 MHz [21]. The reason is that an off-the-shelf small antenna that works along the frequency limits of the Band III was found, as explained in Section 3.3.

Regarding the upper limit of the frequency range to be investigated, the ISM bands above 6 GHz have been also discarded. Such high frequencies would require the use of special substrate materials for the design of printed circuit boards (PCBs) [11] that connect the antenna to the communication modems. This would rapidly increase the cost of the wireless approach, due to the large amount of antennas that must be installed. Finally, a frequency range from 100 MHz until 5.9 GHz has been adopted for the investigation. This range is referred to as "complete frequency range" in the remaining work.

Table 3.1: ISM bands

From	To
6.765 kHz	6.795 kHz
13.553 kHz	13.567 kHz
26.957 kHz	27.283 kHz
40.66 MHz	40.70 MHz
433.05 MHz	434.79 MHz
902 MHz	928 MHz
2.400 GHz	2.500 GHz
5.725 GHz	5.875 GHz
24 GHz	24.25 GHz
61 GHz	61.5 GHz
122 GHz	123 GHz
244 GHz	246 GHz

The antennas under test were not measured in the complete frequency range, because they cannot cover this entire frequency range. The antennas are designed to operate properly only in a particular narrower frequency range. Therefore, the complete frequency range was divided into several narrower sub-ranges, in order to simplify the comparison between antennas. Then, the antennas under test can be grouped and compared first only within the frequency sub-ranges to which they belong. One antenna per group is chosen, so that the performance comparison between the different frequency sub-ranges can be more easily realized with a reduced number of antennas.

Most of the antennas employed were off-the-shelf evaluation boards, in order to investigate several antennas in different frequency sub-ranges in a reasonable time. Most of these antennas are also designed to operate in ISM bands. Therefore, the ISM bands have been used as reference for the determination of the frequency sub-ranges in this work. Each frequency sub-range has a width of 600 MHz and includes one of the ISM bands, and does not overlap with the remaining sub-ranges. The available ISM band in each frequency sub-range of interest and the limits of the four defined frequency sub-ranges are listed in Table 3.2.

Table 3.2: ISM bands and chosen frequency sub-ranges between 100 MHz and 5.9 GHz.

Nr	ISM Band		Freq. sub-range	
	From	To	From	To
1	433.05 MHz	434.39 MHz	100 MHz	700 MHz
2	902 MHz	928 MHz	700 MHz	1300 MHz
3	2400 MHz	2500 MHz	2100 MHz	2700 MHz
4	5725 MHz	5875 MHz	5300 MHz	5900 MHz

3.2 Investigated Antenna Types

All the antennas under test are either integrated in a PCB or are mounted on it. Due to the little height of the available free space inside the battery, antennas with a vertical orientation (perpendicular to the emulator floor) were not taken into account. Besides, antennas with a high directivity were also not considered, since the antennas are expected to offer good performance in every position inside the battery. The following types of antennas have been tested:

Chip antenna: This type of antenna is made of a ceramic material and it is a surface mount device. It is a very small size and compact solution [47]. Even at frequencies below 500 MHz, there are off-the-shelf prismatic models whose largest dimension is smaller than 10 mm. The chip antenna also offers the advantage of being less sensitive than other antenna types (e.g. the planar antennas) to the environment and objects in the neighborhood [42]. Due to its reduced size, this antenna works as an electrically small antenna, especially for the lowest frequency sub-ranges. Consequently, its bandwidth is typically narrower than for other alternatives.

Planar antenna: It is also normally referred to as PCB, low profile, or trace antenna. This type of antenna is currently widely employed for applications such as mobile communications and wireless networks [95]. The planar antenna consists of a radiating trace with a particular form and length on a PCB. Its resonant frequencies would depend on the trace length and how the antenna is fed. In the case of an ideal dipole antenna, its first resonant frequency is at $\frac{2 \cdot c}{L_\mathrm{a}}$, L_a being the length of the antenna and c the speed of the light. In an ideal monopole antenna, the first resonant frequency is at $\frac{4 \cdot c}{L_\mathrm{a}}$ [38]. There are other planar antenna designs, such as the patch antenna or the planar inverted-F antenna (PIFA). From all of them, the PIFA is the most appropriate design to take optimally advantage of the available space on the PCB. A monopole would enlarge the PCB by some centimeters, because its longest trace is perpendicular to one ground plane edge. On the other hand, the PIFA has its longest trace parallel to one PCB edge, so the increase in size of the PCB is smaller. Compared to the patch antenna, the PIFA has a length of one-quarter wavelength, instead of one-half as the patch. Compared to the chip antenna, the PIFA is typically cheaper, since it needs only an extension of the PCB and does not need to be mounted.
For frequencies above 2 GHz, the required antenna trace length is below 3.75 cm, which appears to be a feasible value for the employment inside

batteries. For frequencies below 1.5 GHz, the required length for a PIFA would be too big for the proposed system. However, it can be even employed for the second frequency sub-range by means of sub-optimal alternative designs, as for example the loaded stub antenna [92, 93].

Helix antenna: An alternative wired design to the typical monopole or dipole is the helix antenna (also known as helical antenna). It employs a helix as antenna, i.e. the wire is wound in a helical form. This type of antenna has a much smaller size than the monopole for the same resonant frequency. Helix antennas have a similar pattern as monopole antennas when the spacing between turns and the helix diameter is much smaller than the wavelength at which they resonate [47]. The propagation properties of these antennas have been studied by different authors, such as Kraus [38], and Carr and Hippisley [9]. However, in both cases the authors assumed that the antenna ground plane dimensions must be at least in the order of the wavelength. This would, however, not be the case inside a battery. Since helix antennas are considered as an alternative for the two lowest frequency sub-ranges (up to 1300 MHz), there would not be enough room for such a ground plane on a Li-ion cell. That means that they operate as electrically small antennas and the typical equations applied for the parametrization of these antennas do not describe the characteristics (such as resonant frequency) of the employed helix antennas very accurately. Compared to the PIFA, the helix antenna is more expensive, since it needs to be mounted and the helix form must be manufactured. Therefore, it is not considered for the two highest frequency sub-ranges, where the PIFA already offers a good performance at a lower price and with a smaller size.

3.3 Evaluated Antennas

The antennas under test are listed in Table 3.3. In the remaining work, the numbering of the antennas refers to this table. The antennas that belong to the CC-Antenna-Development Kit [92] from Texas Instruments originally included a matching network, which was not designed for the particular environment of the battery emulator, but for operating practically in free space. Therefore, the matching network components were removed and by-passed. The Black Chip (antenna # 8) from Laird Technologies is a tri-band antenna designed for standards such as IEEE 802.11 and Bluetooth [15]. It can operate in the two highest frequency sub-ranges. The two remaining chip antennas (# 3 and # 6) are manufactured by Linx Technologies Inc. Finally, since an appropriate off-the-shelf PIFA for the 5 GHz – 6 GHz band was not found, it was designed and fabricated at IIIT. The fabrication process is explained in Section 3.5.

Table 3.3: Evaluated antennas

Nr.	Antenna Type	Frequency sub-range [MHz]	Commercial name
1	Double helix[a]	100 - 700	CC-Antenna-Dev. Kit, #14
2	Helix	100 - 700	CC-Antenna-Dev. Kit, #13
3	Chip 433 MHz	100 - 700	AEK-433-USP-ND
4	Helix	700 - 1300	CC-Antenna-Dev. Kit, #11
5	Planar Loaded Stub	700 - 1300	CC-Antenna-Dev. Kit, #10
6	Chip 916 MHz	700 - 1300	AEK-916-USP-ND
7	PIFA	2100 - 2700	CC-Antenna-Dev. Kit, #7
8	Chip Tri-Band	2100 - 2700 5300 - 5900	Eval Board Tri-Band Ant.
9	PIFA	5300 - 5900	Self-Manufactured

[a] The double helix antenna #1 and the Helix antenna #2 employ actually the same off-the-shelf helix, but the antenna #1 has two connected in series.

3.4 Pre-Selection of the Antennas

3.4.1 Performance Comparison by means of the CTF

As explained in Section 2.4, the CTF between antennas in our emulator can be obtained by measuring the S-parameters between the antennas with a VNA. The measured CTFs between antennas of the same type are employed to compare the performance of the different antenna types. Three important aspects of the CTFs must be analyzed: Channel attenuation, frequency selectivity, and position dependence.

Attenuation: The channel attenuation is the inverse of the absolute value of the CTF. A low attenuation is desired, in order to increase the receiving power. In that way, the required transmitting power for a desired bit error rate (BER) at the receiver is reduced [72].

Frequency selectivity: A frequency-selective channel is one whose frequency response is not flat, in contrast to a flat channel. Highly frequency-selective channels present a CTF which strongly varies with the frequency, normally having a high density of notch and resonant frequencies. Such channels produce a time-spreading of the transmitted signal, which can degrade the receiver demodulation performance [82]. Therefore, a more sophisticated (and expensive) receiver would be required.

Position dependence: The position dependence of the channel can be determined by comparing the CTF between two antennas while varying their positions. A low position dependence is desired, in order to simplify the design of the communication system. Otherwise, it could happen, that for a particular operating frequency range, one slave presents a "good" CTF (low attenuation, almost flat response) with the master, but another slave observes a notch as CTF, i.e. high attenuation and high frequency selectivity. Then, a very robust and sophisticated receiver would be required, in order to ensure the proper operation of all the slaves. But in case of low position dependency, it would be possible to find a frequency range in which all the modems present a good CTF with the master, and then, a simple receiver would be sufficient.

Similarity factor

In order to analyze the position dependence analytically, a similarity factor between the CTFs is employed. The similarity factor indicates how similar two functions are, in this case two CTFs. This factor is calculated as the maximum of the cross-correlation function (XCF) between a reference function and a second function under analysis, and it is normalized to the square root of the energy of both functions [18], i.e. the CTFs. The functions under consideration are vectors, i.e. they are in the discrete domain, because the S-parameters are saved as a matrix of points by the VNA: One vector for the frequency values, plus one vector for each S-parameter. The similarity factor (SF) between two discrete vectors X_{ref} and X_i is calculated as follows:

$$\text{SF}_{X_{\text{ref}}X_i} = \max\left(\frac{|X_{\text{ref}} \star X_i|}{\sqrt{E_{X_{\text{ref}}} \cdot E_{X_i}}}\right) \cdot 100\%, \tag{3.1}$$

where X_{ref} is the reference function[1], X_i is the function to be compared, and $E_{X_{\text{ref}}}$ and E_{X_i} are the energies of both functions. The energy of a vector X is calculated as:

$$E_X = \sum_{n=0}^{N-1} |X[n]|^2, \tag{3.2}$$

where N is the number of points of the vector X. Due to the energy normalization, the maximum similarity factor value is 100% (when both functions have exactly the same form) and the minimum is 0% (completely different). Additionally, since the functions are normalized to the energy, the similarity factor is independent of an amplitude offset between the functions. And as the maximum value of the XCF is employed, a frequency offset between the functions also plays practically no role.

[1] When the similarity factor between two CTFs is calculated, the values of the CTFs (X_{ref} and X_i in (3.1)) are always in linear scale, not in dB.

3.4.2 Comparison Procedure

A comparison procedure was developed. First, all the antenna types were measured several times and in different positions inside the small channel emulator. Their performance was analyzed following the three important aspects of the CTF described in Section 3.4.1. The antennas with the best performance were pre-selected to continue the investigation. Then, due to time constrains, not all the antennas under test were measured in the big channel emulator, but only the pre-selected ones[2]. The target for the measurements in the big channel emulator was to prove if the characteristics of the performance of the pre-selected antennas are still valid when the battery dimensions increase.

The following steps were repeated for each type of evaluated antenna in the small channel emulator. First, two antennas of the same type were placed inside the small emulator. One antenna was always located in the center of the emulator. The position of the second one was rotated along four different pre-established positions. The CTFs between both antennas were measured for each position of the second antenna. With the big emulator the process was similar but, as already mentioned, with a smaller number of antennas.

Fig. 3.1 presents a setup with the different positions where the antennas were placed. Table 3.4 depicts the exact coordinates of each position for both emulators.

3.4.3 CTF Measurements in the Small Channel Emulator

The results of the CTF measurements in the small channel emulator[3] are depicted in Fig. 3.2, 3.3, 3.4, and 3.5 for the first, second, third and fourth frequency sub-ranges, respectively. Fig. 3.6 presents the results of the similarity factor for each type of antenna. For the similarity factor

[2] In [107], all the antennas have been also measured in the big emulator. But that was before the emulator enhancement presented in [102]. After that, only the pre-selected antennas were measured in the big emulator.

[3] The results that appear in this work are not the same as in [107]. The channel emulators have been later modified [102] and the measurement results presented in this work were obtained with the improved channel emulators.

analysis, the CTF_{A1B1} was always used as reference, and the similarity factors between this CTF and the three remaining ones were calculated.

Regarding the first frequency sub-range, the CTF plots in Fig. 3.2 show that the form of the four measured CTFs with every antenna type are very similar. This is also verified by the similarity factor values, which are always above 80%. This proves a low position dependence of the CTF form in the lowest frequency sub-range in the small emulator.

Comparing the performance of the three antennas in the lowest frequency sub-range, the chip antenna presents higher attenuation than the helix antennas (around 20 dB higher). Additionally, its bandwidth is narrower. The double helix antenna has a better performance than the other

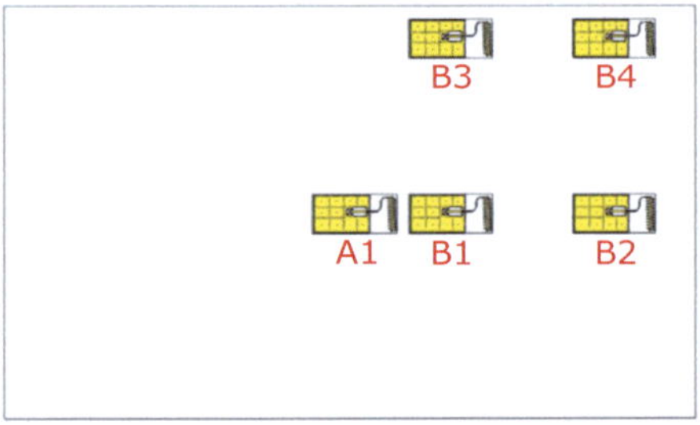

Figure 3.1: Setup of the positions in which the antennas were placed for the experiment described in Section 3.4. The position of the first and fixed antenna is A1. The positions of the second one are represented by B1, B2, B3, and B4 (Positions 1, 2, 3 and 4, respectively).

Table 3.4: Antenna positions, x is the direction parallel to the L, and y parallel to the W. The emulator center is denoted by the (0,0)-coordinate. The dimensions of L and W depend on the employed emulator and they are listed in Table 2.1.

Antenna	Position in Emulator	
	x	y
A	0	0
B1	L/8+10	0
B2	L/8+L/4+10 mm	0
B3	L/8+10	W/2-35mm
B4	L/8+L/4+10	W/2-35mm

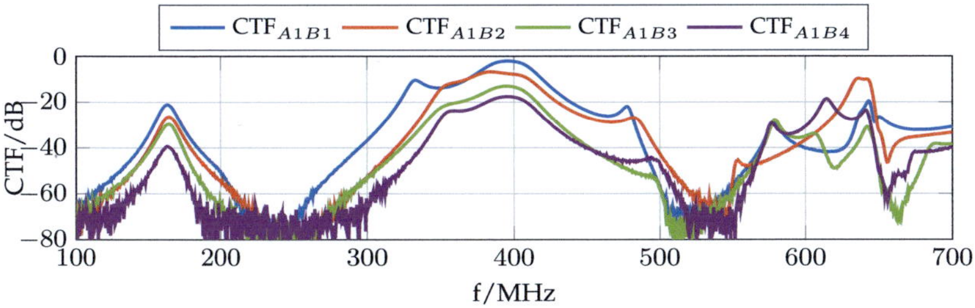

(a) Antenna #1: double helix

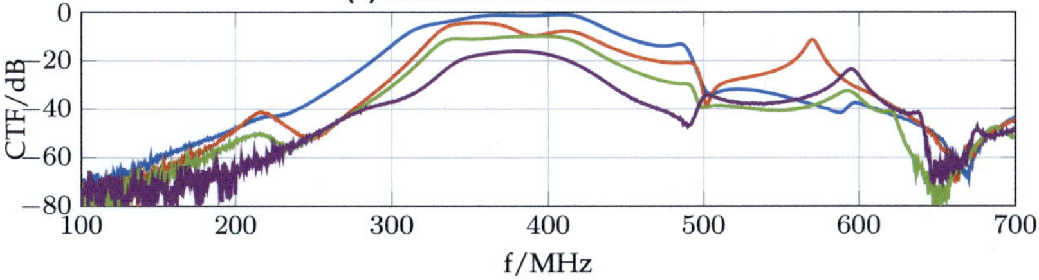

(b) Antenna #2: helix

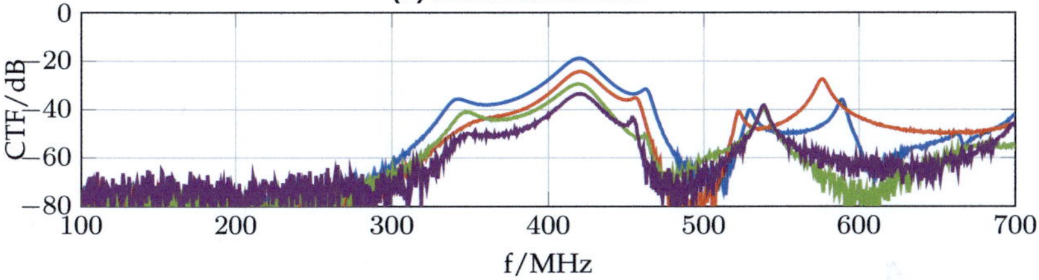

(c) Antenna #3: chip (433 MHz)

Figure 3.2: CTF between one fixed antenna in the center and a second antenna at 4 different positions, frequency range: 100 MHz - 700 MHz. Measurements were made in the small emulator.

two antenna types at frequencies below 200 MHz, since it was originally designed for a central frequency of 169 MHz [93]. Around 400 MHz, both helix antenna alternatives have similar attenuation, but the single helix offers a slightly larger bandwidth.

In the remaining frequency sub-ranges, the antenna position has a much higher influence on the CTF. Most of the similarity factor values are below 50%. In every measured case, the position dependence of the CTF increases when the frequency rises. In the second frequency sub-range, the helix and planar antennas have similar performance, i.e. similar similarity factors and attenuation. The chip antenna type still suffers from much higher attenuation.

The performance of the chip antennas clearly improves at higher frequencies from the third frequency sub-range. The reason is that the ratio between wavelength and antenna size is considerably smaller than in the two lowest frequency sub-ranges, i.e the chip antenna is less electrically small. Above 5 GHz, the chip antenna has even a similar attenuation as

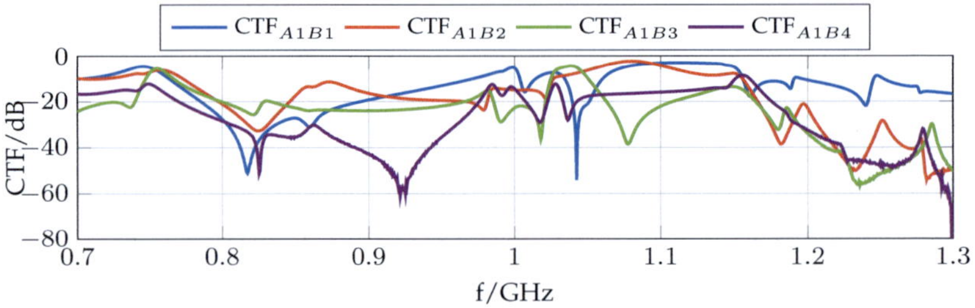

(a) Antenna #4: helix

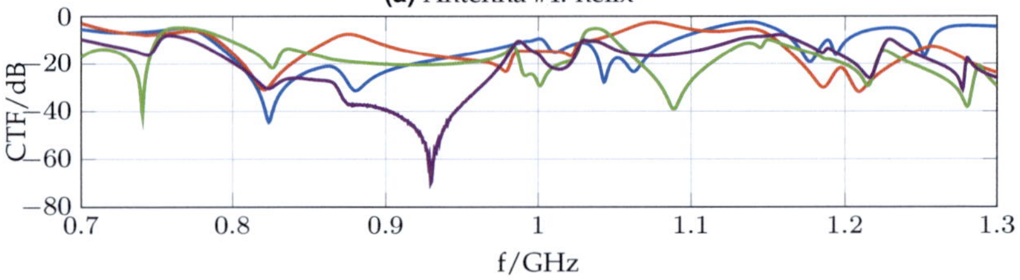

(b) Antenna #5: planar loaded stub

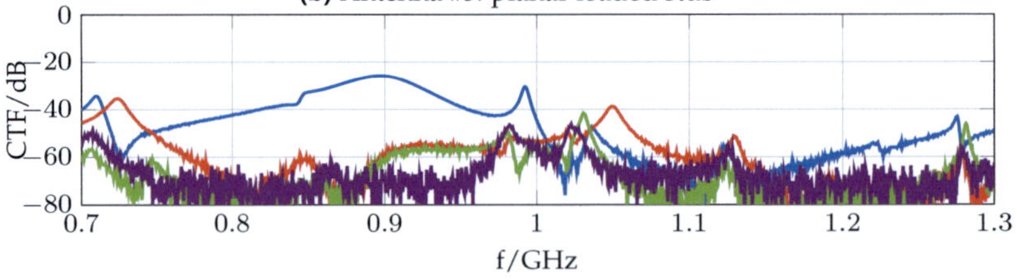

(c) Antenna #6: chip (916 MHz)

Figure 3.3: CTF between one fixed antenna in the center and a second antenna at 4 different positions, frequency range: 700 MHz - 1300 MHz. Measurements were made in the small emulator.

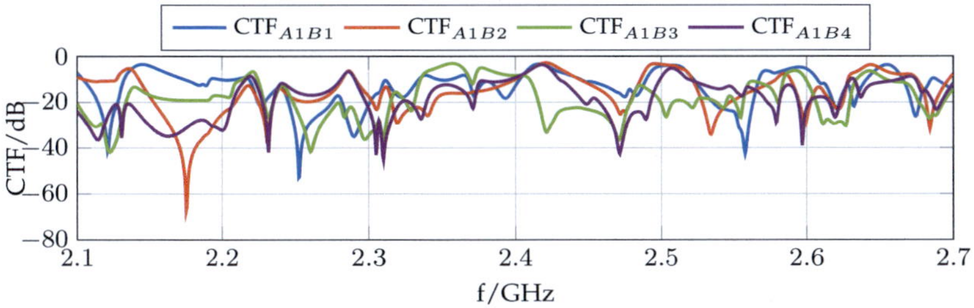

(a) Antenna #7: PIFA

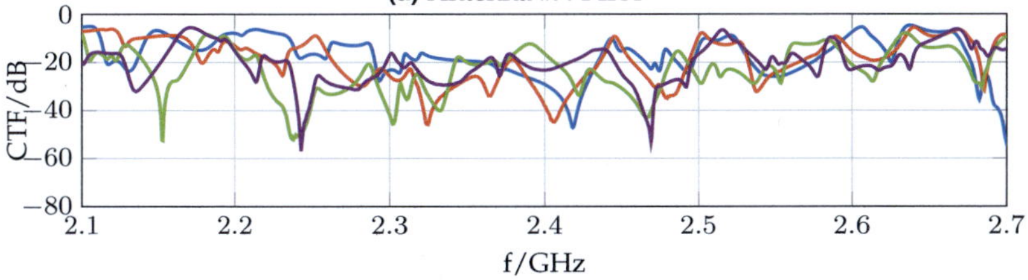

(b) Antenna #8: chip tri-band

Figure 3.4: CTF between one fixed antenna in the center and a second antenna at 4 different positions, frequency ranges: 2100 MHz - 2700 MHz. Measurements were made in the small emulator

the PIFA. The last antenna type offers low attenuation both in the third and fourth frequency sub-ranges.

Comparing the CTF plots of all the antennas, it can be observed that the higher the frequency, the larger is the amount of resonant and notch frequencies. As already explained in Section 2.2, the battery emulator has many similarities with a rectangular cavity resonator. Consequently, the higher the frequency, the higher is the density of resonant frequencies. This increases the frequency selectivity of the channel.

3.4.4 Pre-Selected Antennas

Summarizing the above results, the CTF at lower frequencies has a very interesting characteristic: Smaller dependency on the antenna position. This would simplify the design of a communication system for many reasons, as it will be explained in Chapter 6.

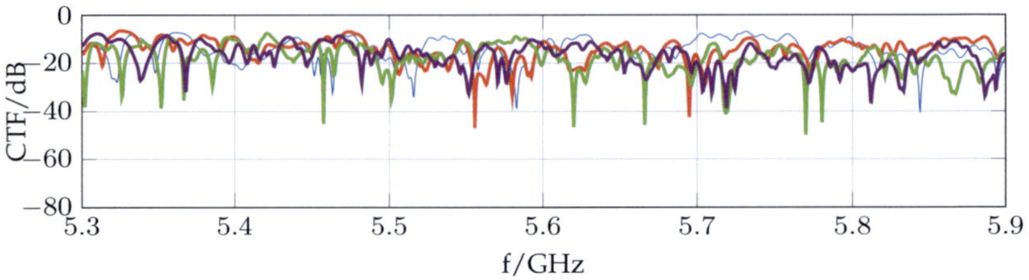

(a) Antenna #8: chip tri-band

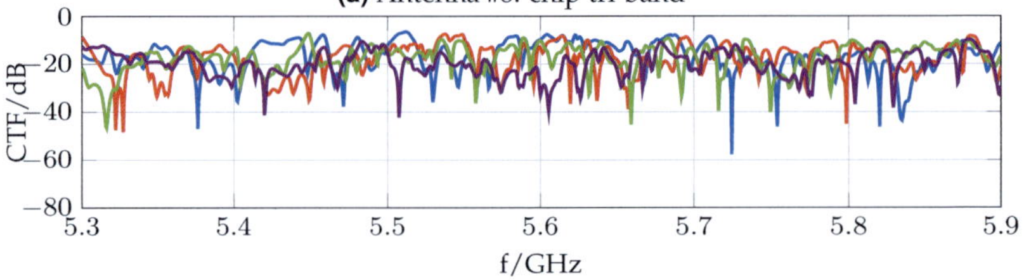

(b) Antenna #9: PIFA

Figure 3.5: CTF between one fixed antenna in the center and a second antenna at 4 different positions, frequency range: 5200 MHz - 5800 MHz. Measurements were made in the small emulator.

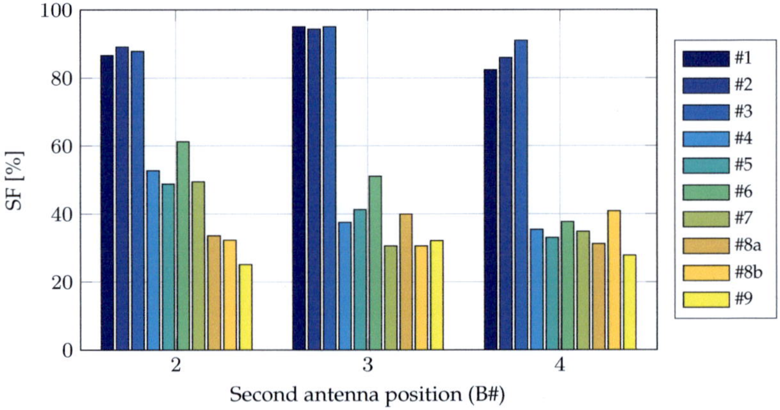

Figure 3.6: Similarity factor as function of the second antenna position and the number of antenna (written in the legend). Measurements were made in the small emulator.

Among the antennas tested in the first sub-range, the single helix antenna #2 offered a similar performance as the double helix antenna #1, but it is considerably smaller and obviously cheaper. The chip antennas are neither considered for the first frequency sub-range nor for the second one, due to the higher attenuation. The helix antenna #4 for the second sub-range is not a better alternative than the antenna #2, since it is only slightly smaller and it does not offer the advantage of the low CTF's dependence on the antenna position.

The planar antennas have very low attenuation at the resonant frequencies for the two highest frequency sub-ranges. They also offer further advantages, such as being smaller than the helix antennas and also cheaper, since only a small trace of PCB is required. However, they have the disadvantage of higher position dependence. Compared to chip antennas, the last ones need to be mounted (i.e. placed and welded) on the PCB. Therefore, the manufacturing process would be more expensive than for the planar antennas.

Based on the results and conclusions exposed before, it was decided to continue with the next step (the measurements in the big emulator) analyzing only three different alternatives:

1. Helix antenna #2 for the first frequency sub-range
2. PIFA #7 for the third frequency sub-range
3. PIFA #9 for the fourth frequency sub-range

3.4.5 CTF Measurements in the Big Channel Emulator

The results of the CTF measurements with the three pre-selected antennas are depicted in Fig. 3.3. The similarity factors appear in Fig. 3.8.

Comparing these results with the ones in the small emulator, it is clear that the amount of resonant frequencies in the CTFs increases, while the similarity factors decrease. This can be explained by the resonant behavior of the channel emulator. In a rectangular cavity resonator, the resonant frequencies can be obtained as [71]

$$f_{mnl} = \frac{c}{2\sqrt{\mu_r \epsilon_r}} \sqrt{\left(\frac{m}{x}\right)^2 + \left(\frac{n}{y}\right)^2 + \left(\frac{l}{z}\right)^2}, \tag{3.3}$$

where the indices m, n, and l indicate the resonant mode numbers in the directions x (width), y (length) and z (height), respectively; c is the speed of light in vacuum; and μ_r and ϵ_r are relative permeability and relative permittivity of the cavity filling, respectively. μ_r and ϵ_r can be approximated as one, since the cavity is filled with air.

From (3.3) is clear that increasing the emulator size results in a more frequency-selective channel, due to the higher amount of resonant frequencies. The numbers of resonant frequencies in each frequency sub-

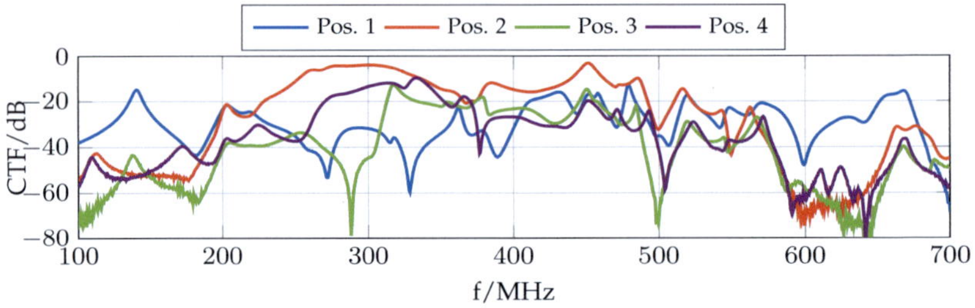

(a) Antenna #2: Helix for the first frequency sub-range

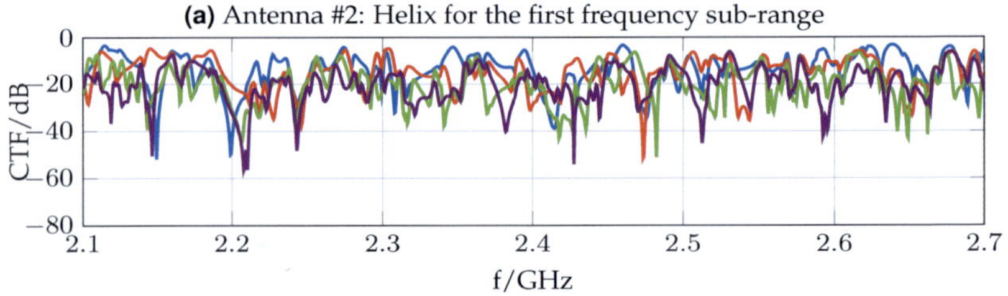

(b) Antenna #7: PIFA for the third frequency sub-range

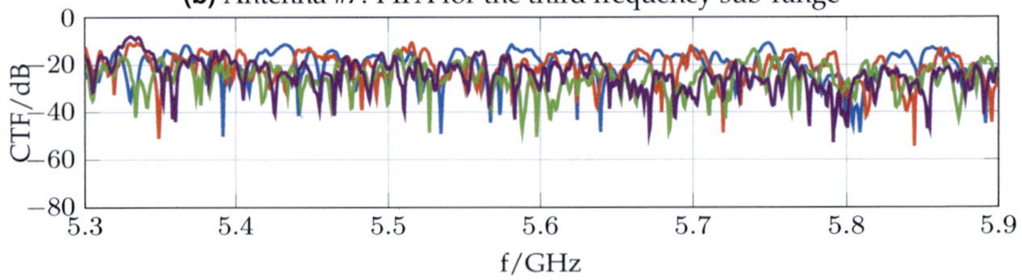

(c) Antenna #9: PIFA for the fourth frequency sub-range

Figure 3.7: CTF between one fixed antenna in the center and a second antenna at 4 different positions. Measurements were made in the big emulator.

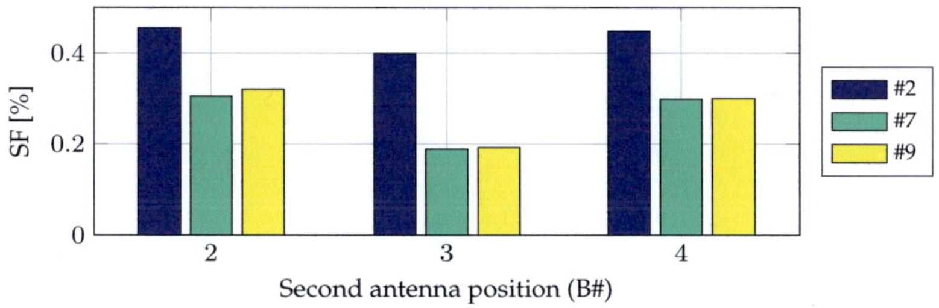

Figure 3.8: Similarity factor as function of the second antenna position and the number of antenna. Measurements were made in the big emulator.

range are calculated for rectangular cavity resonators with the same dimensions as the emulators and presented in Table 3.5.

Taking into account the last results, it can be stated that the concept of "low" or "high" frequency actually depends on the relationship between the emulator dimensions and the wavelength. Therefore, the correct statements regarding the frequency selectivity and position dependency of the CTF are:

- The bigger the ratio between the emulator dimensions and the wavelength, the more frequency-selective is the CTF, because the number of resonant modes increases.
- The bigger the number of resonant modes, the more position dependent is the CTF.
- Observing the measurements with the big emulator, it can be pointed out that if the battery size increases, the helix antenna in the first frequency sub-range loses its advantage of low position dependence of the CTF. Then, the PIFA becomes the best alternative, due to its lower cost and smaller size.

The influence on the CTF of the several system parameters, such as battery size, number of antennas, etc., is expanded in Section 5.1.1 by means of electromagnetic simulations.

Table 3.5: Amount of resonant frequencies in the empty emulators (assuming ideal rectangular cavity resonators) for each frequency sub-range

Freq. sub-range #	Small Emulator	Big Emulator
1	4	24
2	15	65
3	37	167
4	178	778

3.5 Manufacturing of Pre-Selected Antennas

As mentioned in Section 3.1, the antennas measured during the previous test are actually off-the-shelf evaluation boards, with the exception of the PIFA for the fourth frequency sub-range. In the next chapter, the selected antenna models for software-based 3D electromagnetic simulations are presented. With the target of reducing the complexity of these simulation models and to achieve a better matching with the measurements, similar antennas as the pre-selected ones have been designed and fabricated in the laboratory of the IIIT at KIT. In this way, the exact geometry of the antennas is known and the simulation models can be developed more accurately. Additionally, the self-designed antennas can be simplified, for example, employing a single layer ground plane (instead of a double layer as in the evaluation boards). In this way, also the vias that connect both layers can be avoided. These simplifications reduce the simulation times.

All the manufactured antennas employed FR4 glass epoxy as PCB substrate, the same as the used off-the-shelf ones. FR4 is not the best choice for high frequencies, such as the fourth frequency sub-range, because of the degradation of the material properties when the frequency increases [11]. However, the PCB of the off-the-shelf evaluation board of the chip antenna #8 also employs FR4 as substrate. The use of a better substrate for high frequencies would increase the cost of the antennas, making this alternative senseless. Therefore, FR4 was also employed for the self-manufacturing of the PIFA #9.

In the case of the helix antenna, the same off-the-shelf helix was used. The PCB was designed according to the dimensions of the antenna #2.

However, after having manufactured and measured the S-parameters of several units of this antenna in the small channel emulator, it was observed that the measurement results were not accurately reproducible, i.e. the different antennas have clearly different S-parameters. The results were very sensitive to the manufacturing process. E.g. modifying the soldering between the helix and PCB had an important impact on the measurement results. However, the same issue was not observed with the off-the-shelf helix antenna #2. With this antenna the results were much better reproducible. Therefore, for the investigation of the channel and the comparison between measurement and simulations, it was decided to continue working with the off-the-shelf helix antenna #2, instead of the customized antenna. Nevertheless, this customized helix antenna can be used when the channel is not the topic under study. For example, when the antennas are used to test the communication prototypes, as in Chapter 7.

Regarding the PIFAs for the third and fourth frequency sub-ranges, the equations presented in [37] were employed as first reference for the designs. Then, the final dimensions were adjusted by means of electromagnetic simulations (more details in Chapter 4). In the case of the PIFA for the third frequency sub-range, the interchange between the manufactured antennas did not have a significant impact on the measurement results in the emulator. The reason is presumably that the manufacturing process is simpler than the one of the helix antennas, since for the PIFAs only the SMA connector must be mounted, but the antenna is already part of the manufactured PCB. Therefore, the customized antennas for this frequency sub-range were employed for the remaining work. However, when measuring the PIFA for the fourth frequency sub-range, it was again observed that the measurement results were hardly reproducible, even without interchanging antennas. This is explained in Section 3.7.1. The customized antennas and their physical dimensions are shown in Fig. 3.9.

3.6 Consequences of the Antenna Orientation

As explained in Section 2.3 the antennas are placed parallel to the emulator floor, in order to take better advantage of the available free space in the emulator (or in a real battery). Then, as consequence of this orientation and the type of the employed antennas, the electrical field radiated by the antennas also has an orientation parallel to the emulator floor.

If the channel emulator is analyzed as a rectangular cavity resonator, the resonant frequencies inside the emulator can be obtained with (3.3). In the case of the manufactured channel emulators it can be stated that $z << x < y$, since $H << W < L$. Therefore, the first (lowest) resonant frequencies inside the emulator have the index $l = 0$, and hence, the electrical field orientation of the propagated wave should be perpendicular to the emulator floor. Resonant frequencies with $l > 0$ (electrical field parallel to the emulator floor) appear from a particular frequency value, denoted as $f_{\min(l>0)}$ in this work. The value of $f_{\min(l>0)}$ can be calculated as

$$f_{\min(l>0)} = \frac{c}{2}\sqrt{\left(\frac{1}{L}\right)^2 + \left(\frac{1}{H}\right)^2} \tag{3.4}$$

since $L > W$, otherwise L must be replaced by W [102, 105]. Considering the dimensions of the channel emulators, $f_{\min(l>0)} = 3.0061$ GHz for the

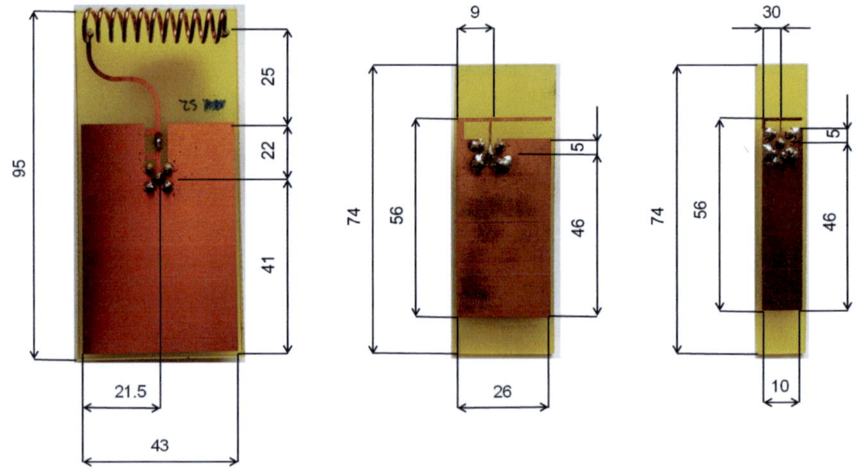

Figure 3.9: Photo of the customized antennas, from left to right: Helix antenna, PIFA for the third frequency sub-range, and PIFA for the fourth frequency sub-range. The dimensions of the helical wire of the helix antenna can be found in the corresponding datasheet [86].

small channel emulator and $f_{\min(l>0)} = 3.0017\,\text{GHz}$ for the big channel emulator.

No wave propagation is allowed for the antennas in the first three frequency sub-ranges, since these sub-ranges are below $f_{\min(l>0)}$, and the polarization of the electrical field is parallel to the emulator floor. However, good coupling, i.e. a channel with low attenuation between antennas is still possible, due to the short distances between the antennas, as well as the presence of reflective objects in the near field region of the antennas, such as the metallic walls, floor, and top of the emulator. The channel attenuation around the resonant frequencies is even much smaller than in typical commercial wireless systems such as wireless local area networks (WLANs) or mobile communications.

For frequencies above $f_{\min(l>0)}$, the density of resonant frequencies (the amount of resonant frequencies in a particular frequency range) increases and wave propagation is also allowed in the direction of the polarization of the electrical field produced by the antennas. Consequently, the channel is much more frequency-selective, as proved by the values of Table 3.5. This also increases the delay spread of the channel, i.e. it decreases the channel coherence bandwidth, as it explained in Section 5.5.

3.7 Final Selection of the Antennas

3.7.1 High Frequency Measurements inside the Emulator

To study the low reproducibility of the PIFA properties for the highest frequency sub-range, the following experiment was carried out, first with the customized PIFA for the third frequency sub-range, and then also with the customized PIFA for the fourth frequency sub-range.

1. Two antennas of the same type were placed inside the channel emulator, in the positions A1 and B1, according to Fig. 3.1.
2. The CTF between the antennas was measured.
3. The antenna in the position B1 was rotated by 5°.

The second and third steps were repeated three more times. The CTF was measured in total four times with each PIFA type, for a rotation of 0°,

5°, 10° and 15° of the antenna in the position B1, according to Fig. 3.1. Fig. 3.10(a) and Fig. 3.10(b) depict the measurement results for the PIFA for the third and fourth frequency sub-ranges, respectively. In both cases, the four plotted CTFs are practically overlapping only up to a frequency of around 2.8 GHz. Up to this frequency, a slight rotation of one of the antennas does not play an important role in the CTF. For higher frequencies, the CTF is highly sensitive to the variation of the antenna orientation.

A second experiment was also carried out. This time, a little pressure was applied to the emulator top on the different corners of the emulator, pushing down the top with a hand. In this way, the contact between the

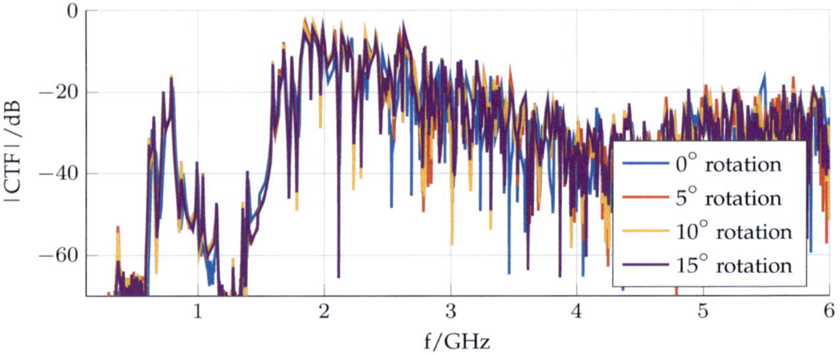

(a) PIFA for the third frequency sub-range

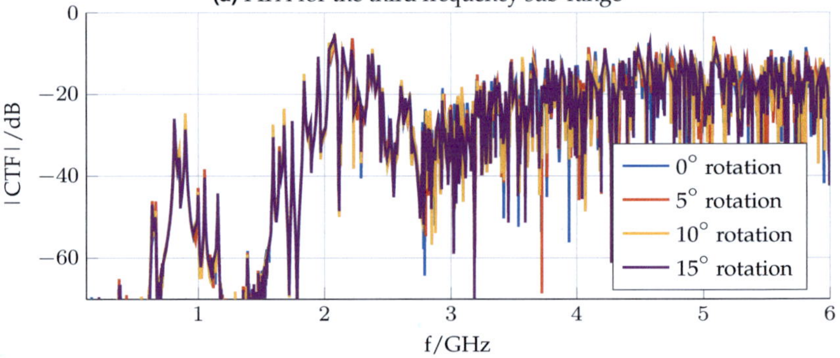

(b) PIFA for the fourth frequency sub-range

Figure 3.10: Effect on the CTF of slightly rotating one of the antennas in the emulator

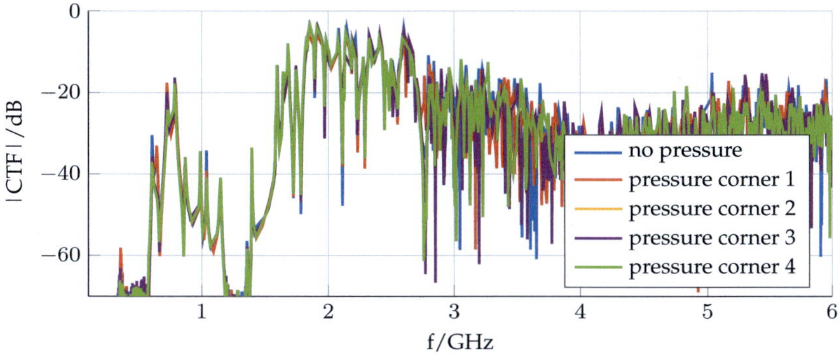

(a) PIFA for the third frequency sub-range

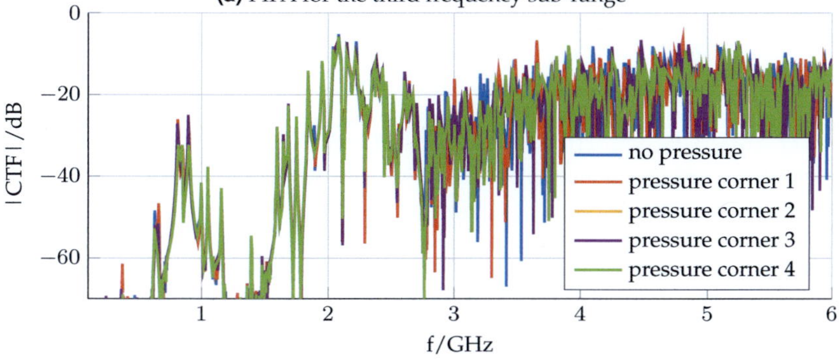

(b) PIFA for the fourth frequency sub-range

Figure 3.11: Effect on the CTF of slightly applying pressure on the different corners of the emulator top by pushing it down with the hand

emulator top and the emulator walls is slightly modified. The CTF was measured in every case. The results can be observed in Fig. 3.11. The conclusion was again the same as with the first experiment: The CTFs are very similar up to a frequency of around 2.8 GHz. For higher frequencies the measured CTFs are completely different.

The reason found for the described situation is again related to the resonant behavior of the channel emulator [102, 105]. At higher frequencies and in particular for frequencies above $f_{\min(l>0)} = 3.0061$ GHz, the density of resonant frequencies increases considerably. Additionally, $f_{\min(l>0)}$ is calculated for an ideal and empty rectangular cavity resonator. However, the presence of the antennas inside the channel emulator also modifies the wave channel characteristics and wave propagation. Then, the variation of different parameters related to the antennas inside the emulator, such as their position, orientation, quantity, geometry, etc., also modifies the resonant modes of the emulator.

The consequence of the explained issue is that the CTF measurements in the channel emulator for frequencies higher than approximately 2.8 GHz are hardly reproducible. Therefore, a comparison of these measurement results with simulation results would be senseless. Hence, in the remaining work, antennas are only simulated up to 2.8 GHz, and the PIFA for the fourth frequency sub-range is not considered for the studies based on simulations.

3.7.2 Selected Antennas

Taking the measurement results and the issues with the antennas for the highest frequency sub-range into account, finally two antenna types are considered for the rest of the analysis in this work:

- The helix antenna for the first frequency sub-range
- The PIFA for the third frequency sub-range.

In the rest of this work, only the customized PIFA will be employed for the third frequency sub-range and it will be referred to only as PIFA. The off-the-shelf helix antenna #2 is employed for the channel measurements and simulations in Chapters 4, 5, and 6. The customized helix antenna is used for the tests of the prototypes in Chapters 7 and 8.

The customized antennas have been already shown. The remaining selected antenna, the off-the-shelf helix antenna, is depicted in Fig. 3.12.

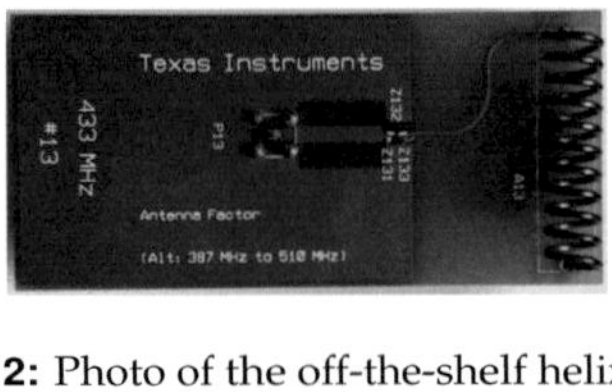

Figure 3.12: Photo of the off-the-shelf helix antenna.

4 3D Electromagnetic Simulation of the Channel

In Section 2.1, the importance of developing a channel emulator, due to the complexity of working with a real automotive battery, has been explained. The channel emulator simplifies significantly the measurement of the CTF between antennas. However, a deeper investigation of the wireless channel in the emulator requires several additional experiments that could not be easily carried out by measurements. For example, if the effect of a bigger amount of antennas inside the emulator must be studied, such a measurement would be much more expensive and would require a long preparation time, since more antennas should be fabricated. The complete measurement of every CTF between master and slaves would also take considerably longer, due to the limited number of ports of the VNA. Another even worse case would be the investigation of the influence of the emulator dimensions, since several new emulators with different dimensions would have to be constructed.

The adopted solution for the described problems was to continue the channel analysis by means of software-based 3D electromagnetic simulations. The software Computer Simulation Technology–Microwave Studio (CST MWS) has been employed. This simulation tool offers a computer-aided design (CAD) tool to recreate the antennas and their complete environment. Among several simulation properties, the boundary conditions must be defined, as well as the type and position of the excitation sources (such as discrete port, waveguide port, etc.). CST MWS is capable of measuring different parameters between the excitation sources similar to a VNA, as for example the S-parameters [10]. It can also provide several additional results, such as the field distribution in the model, near and far fields, etc. Furthermore, simulations also offer the possibility of validating the measurement results.

4.1 Simulation Models

Simulation models of the selected antennas listed in Section 3.7.2 were developed for the software CST MWS. They are shown in Fig. 4.1. There, it can be observed the model of the SMA connector for the antenna feed at the back of the antenna. A waveguide port was employed for feeding the antennas in every simulation, since the coaxial cables employed in the real emulator also work as circular waveguides [91].

The complete simulation model of the testbed always consists of the following devices:

- The channel emulator
- the antennas to be placed
- an extension of each SMA connector

One example of the complete testbed is shown in Fig. 4.2, in this case with five PIFAs in the small emulator. The SMA extension is to simulate the SMA adapters shown in Fig. 2.2(b). The coaxial cables connected between these adapters and the VNA ports must not be simulated, since

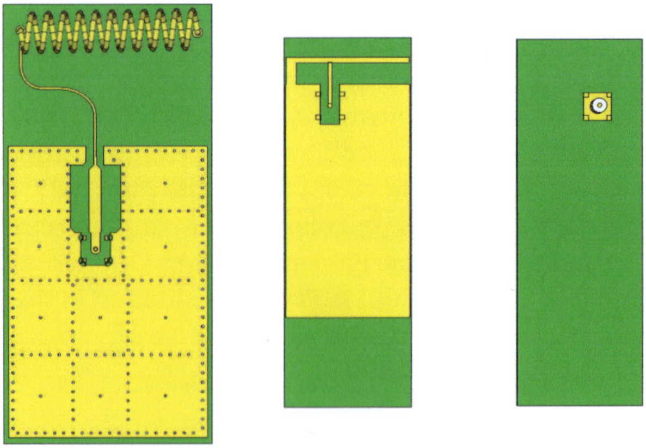

Figure 4.1: Antenna simulation models. Left: Helix antenna #2 for the first frequency sub-interval (front view), center: PIFA (front view), right: PIFA for the third frequency sub-interval (rear view)

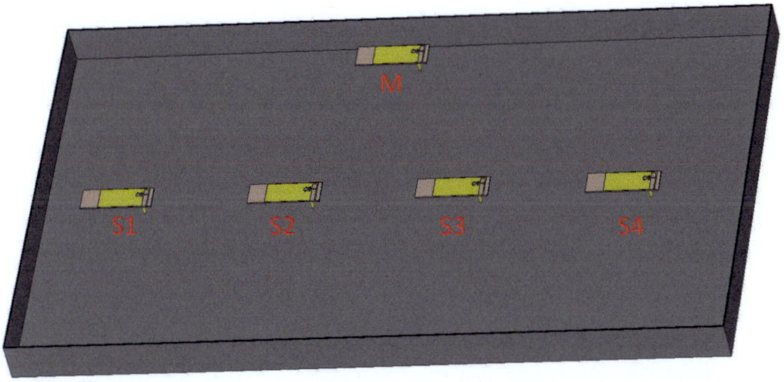

Figure 4.2: Simulation model of the complete model (cut view, parallel to the floor plane)

the VNA is calibrated to measure the S-parameters only from the end of these cables, i.e. at the beginning of the adapter.

The simulation of every complete testbed takes an extremely long time with a commercial desktop computer. Therefore, in order to complete the large amount of simulations required in a reasonable time, a computer cluster of the Karlsruhe Institute of Technology was employed. However, there is a limit established for the maximum duration of the simulations in this cluster. The completion of the simulations with the big emulator without exceeding this duration limit was not achieved and, consequently, only the small emulator was simulated.

4.2 Selection of the Simulation Solver

The software CST Microwave Studio includes many different solver techniques for the electromagnetic simulations. In this work, two solvers were tested: transient (time domain) solver and frequency domain solver. Both of them are general purpose 3D electromagnetic simulators [10] and can calculate the field distribution inside the model, the propagation of near and far fields, and the S-parameters, among others. The software documentation recommends to employ the transient solver for electrically big structures. The frequency domain solver is the recommended alternative in case of working with electrically small components or devices with a high selectivity (high Q-factor). The selection of the solver for the simulation of the complete testbed is not trivial, due to the big difference in size between the antennas and the channel emulator.

The S-parameters of a particular testbed were measured and simulated with both solvers to compare the results in order to choose the most suitable solver for the simulations in the framework of this work. The particular testbed consisted of two PIFAs inside the emulator in the positions A and B2 according to the scheme in Fig. 3.1. Fig. 4.3 shows the results of the CTF between the antennas, i.e. the S_{21} parameter. There, it can be observed that both simulation results are similar in form but shifted in frequency. The frequency offset against the measurement result is bigger with the frequency solver. Additionally, the simulation duration was shorter with the transient solver. Therefore, the last one was employed for the remaining simulations in this work.

4.3 Cross-Validation with Measurements

Several situations were first measured and then simulated in order to validate the results. First, only the antennas, both helix and PIFA, were validated without the emulator. Then, the complete testbed too [102].

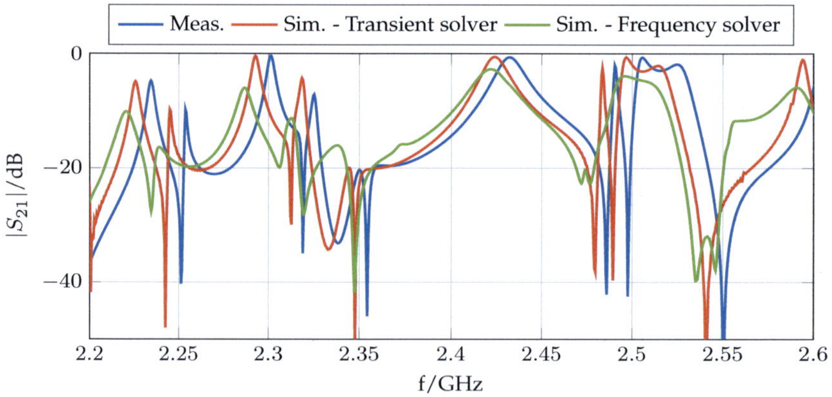

Figure 4.3: Comparison of the results of a measurement with the corresponding simulations with transient and frequency solver

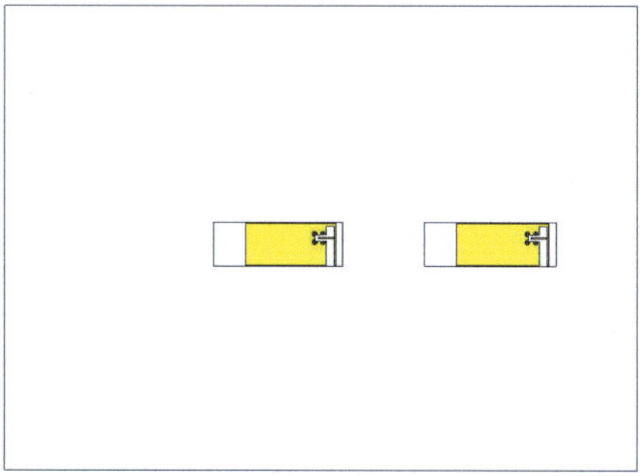

Figure 4.4: Simulation model of two PIFAs above the metallic plate.

4.3.1 Antennas

4.3.1.1 Method

Two experiments were carried out in order to validate the antenna simulation models independent of the emulator model. Both experiments were first set up and measured with a VNA. Then, they were reproduced in the software CST–MWS and simulated.

In the first experiment, the S_{11} of each antenna type under study was measured and simulated to compare the results. However, as already explained in Section 2.3, measuring one isolated antenna of small dimensions can normally lead to inaccurate results. Therefore, the following setup was employed: The antenna under test was placed over a rectangular metallic plate of 290 mm x 390 mm. The antenna feed point was placed in the center of the metallic plate and connected in the same way as the antennas in the testbed, as shown in Fig. 2.2. The antenna was connected to a VNA in the measurement and to the excitation waveguide port in the simulation. In the second experiment, a second antenna (of the same type as the first one) was placed at a distance of 130 mm from the first one, as shown in the simulation model in Fig. 4.4, that corresponds to the case of the PIFA. In this case, S_{21} (that corresponds to the CTF) was the parameter under analysis.

4.3.1.2 Results and Discussion

Custom Made Helix Antenna As explained in Section 3.5, this antenna has a problem with not reproducible properties. Two custom made antennas of this type were measured individually above the metallic plate and the obtained results of S_{11} were clearly different. Then, it is not surprising that the matching between measurement and simulation results is very poor. The comparison of results between simulation and the measurement of one of the customized antennas is shown in Fig. 4.5 for the first experiment and in Fig. 4.6 for the second one. Due to the poor matching between measurements and simulations, the customized helix antenna was not employed for the electromagnetic simulations in the remaining chapters.

Off-the-shelf Helix Antenna: Fig. 4.7 shows the results of the first experiment with this antenna. There, the S_{11} parameter from the measurement and from the simulation are compared in a frequency range from 200 MHz to 2 GHz. Four resonant frequencies can be clearly identified in the measurement result at the following frequency values: 402 MHz, 1038 MHz, 1329 MHz, and 1840 MHz. In the simulation four resonant frequencies appear at similar values: 387.2 MHz, 1030 MHz, 1320 MHz, and 1834 MHz. The first mentioned resonant frequency is the only one included in the first frequency sub-range (the one of interest for the helix antenna). The

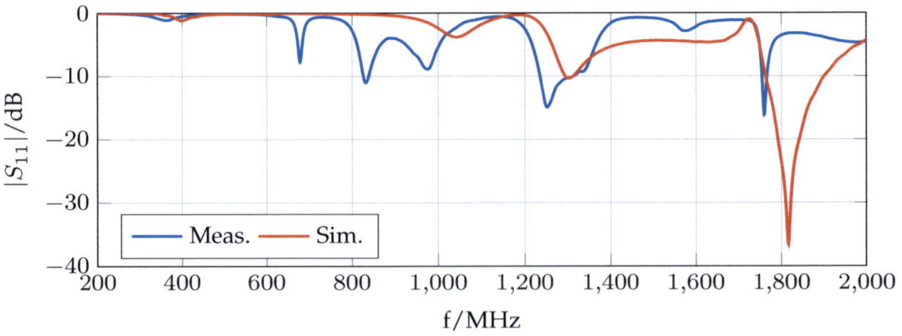

Figure 4.5: S_{11} measurement and simulation results of a customized helix antenna above a metallic plate

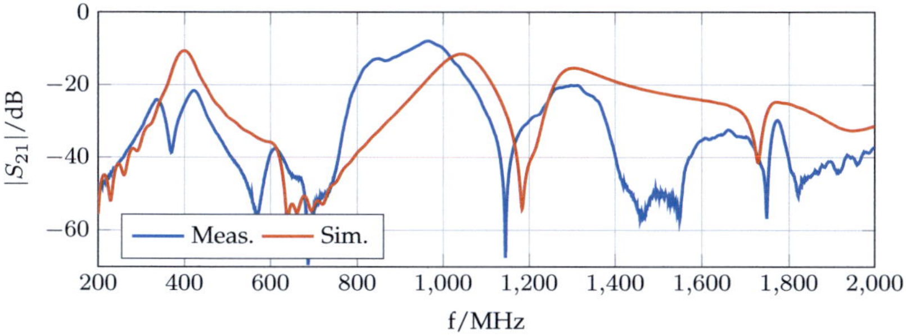

Figure 4.6: S_{21} measurement and simulation results between two customized helix antennas above a metallic plate

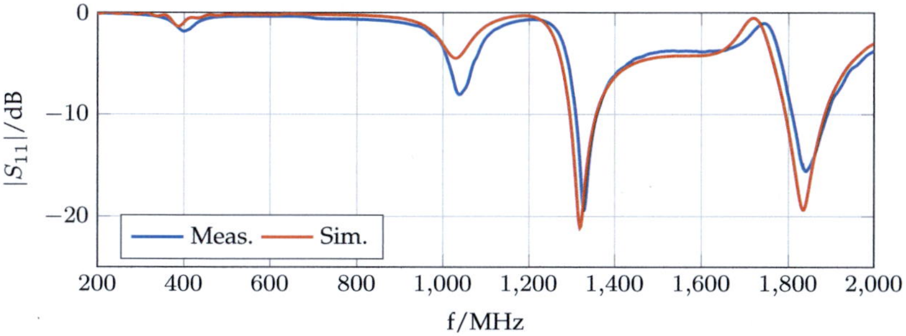

Figure 4.7: S_{11} measurement and simulation results of an off-the-shelf helix antenna above a metallic plate

frequency offset between the measurement and the simulation results is 14.3 MHz, i.e. 3.56% of the measured resonant frequency.

The results of the second experiment are shown in Fig. 4.8. Again, the most significant resonant frequencies, i.e. where S_{21} reaches its highest values, are clearly coincident. Regarding the maximum value of S_{21} in the first frequency sub-range, the measurement and simulation results have an offset of 6.7 MHz in frequency and 2.96 dB in amplitude.

PIFA: The comparison of the measurement and simulation results of the S_{11} parameter in the first experiment is shown in Fig. 4.9. There is only one resonant frequency in the frequency range of interest for this antenna (third frequency sub-range) in both measurement and simulation. The resonant frequency is 2.57 GHz in the measurement and 2.50 GHz in the simulation, i.e. 2.46% of offset respect to the measured value. The different in the magnitude of the S_{11} at the corresponding resonant frequencies is 0.45 dB.

Fig. 4.10 shows the results of the second experiment with the PIFA. A high similarity factor of 91% between measurement and simulation was obtained. However, a high frequency fluctuation in the amplitude of S_{21} is observed in the measurement, but not in the simulation. This is probably caused by reflections at objects located close to the antennas in the real measurement scenario, as for example the VNA. These elements were not included in the simulation. This is, however, not an important issue for this work. The relevant measurements and simulations are the ones of the complete testbed. In these cases, the own emulator acts as a shield, preventing reflections by external objects.

The results with the off-the-shelf helix antenna and the PIFA in this section prove a very good matching between measurements and simulations. The next step is to validate the complete testbed with these two types of antennas. The customized helix is not any longer considered for simulations.

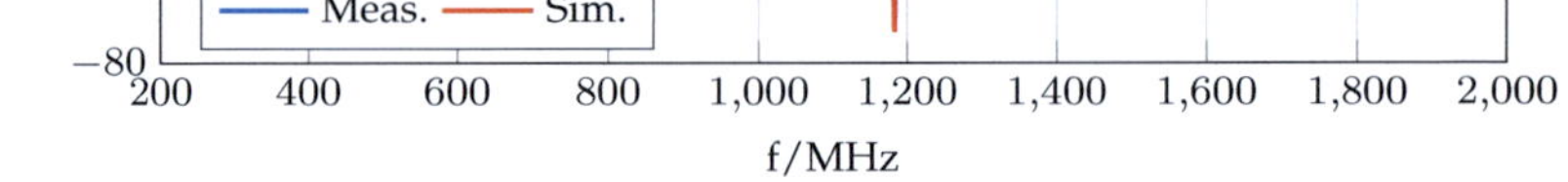

Figure 4.8: S_{21} measurement and simulation results between two self-customized helix antennas above a metallic plate

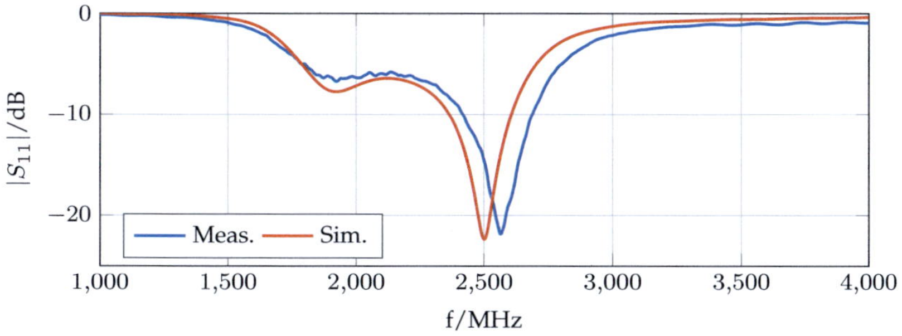

Figure 4.9: S_{11} measurement and simulation results of a customized PIFA above a metallic plate

4.3.2 Complete Testbed

4.3.2.1 Method

The commercial battery employed as reference for the design of the small emulator has 96 cells, divided in eight modules with 12 cells each of them. The modules are uniformly distributed into two parallel levels, one above another. Following the explanation in Section 1.3.3 and considering the case of a module-wise communication, one master antenna should be placed in each level and communicate with the four slave antennas on this level. This situation has been measured and simulated employing the small channel emulator. Five antennas were placed in this emulator. Fig. 4.2 shows the model with PIFAs. The one closest to the wall acts as master, the remaining four placed along one line are the slaves. The exact antenna positions are indicated in Table 4.1. Only the channels between the master and each slave are analyzed, but not the channels between slaves, since the communication between slaves is not required.

The helix antennas were not simulated between 100 MHz and 600 MHz, but between 200 MHz and 500 MHz. This last frequency range includes the widest CTF interval with low and almost constant attenuation. A simulation in the complete frequency sub-range would require a longer time, but it would not bring any additional useful information.

The following parameters were calculated for each pair of measurement and simulation results:

- Similarity Factor: already explained in (3.1)
- Frequency offset (Δf): From the calculation of the XCF in (3.1), the position of the maximum peak of the XCF represents the frequency offset (horizontal offset between the CTF plots) between the measured and simulated CTFs
- Root mean square (RMS) value in decibels of the measured and simulated CTFs: The difference between these two values represents the average attenuation offset in decibels

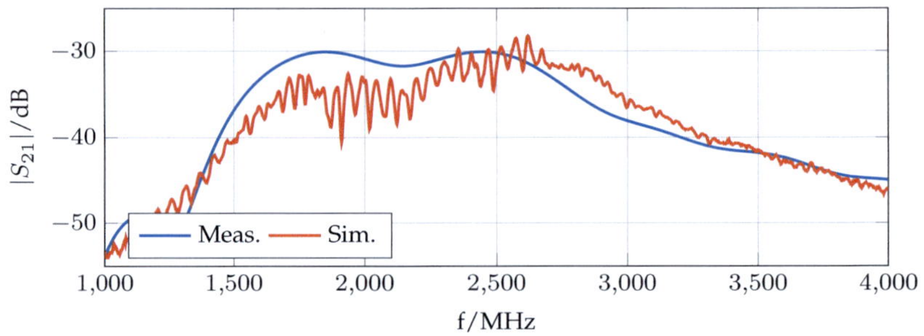

Figure 4.10: S_{21} measurement and simulation results between two customized PIFAs above a metallic plate

Table 4.1: Antenna positions for the scenario described in 4.3.2. The (0,0)-coordinate corresponds to the center of the emulator. L is the length of the emulator. x and y are the coordinates in the directions of the length and the width, respectively.

Antenna	Position in Emulator	
	x	y
M	0	W/2-35mm
S_1	-(L/8+L/4+10 mm)	0
S_2	-(L/8+10 mm)	0
S_3	L/8+10 mm	0
S_4	L/8+L/4+10 mm	0

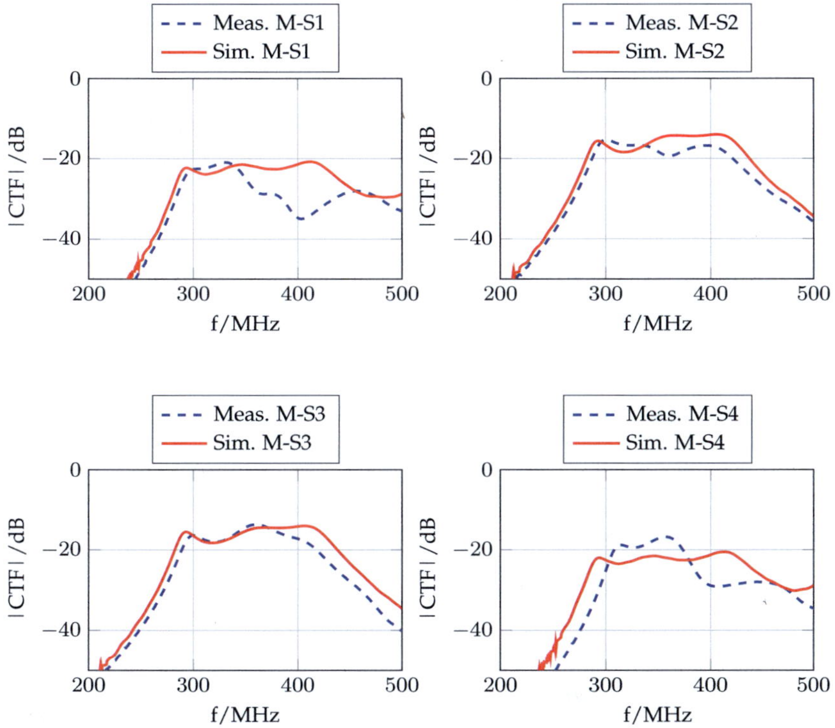

Figure 4.11: Validation of the CTF measured and simulated in the channel emulator with helix antennas

4.3.2.2 Results and Discussion

The CTF results of both measurement and simulation in each position are presented in the Fig. 4.11 for the helix antennas and in Fig. 4.12 for the PIFAs. Additionally, the results of the calculated parameters are depicted in Table 4.2 for the helix antennas and in Table 4.3 for the PIFAs.

Helix Antenna: A good matching between measurements and simulations of the CTF is proven by the high similarity factor values. For frequencies below 350 MHz, the simulation and measurement CTF curves are almost coincident in three out of four slave positions. The two positions closer to the center and to the master antenna have the highest matching. The slaves that are placed symmetrically according to the line segment bisector of the emulator length, and consequently at the same distance of the master, present very similar CTFs. For example, the CTFs M-S2 and M-S3, or M-S1 and M-S4. All the RMS values of attenuation offsets between measurement and simulation of the CTF is always below 3 dB. The frequency offset is relatively high for the position M-S4, but for the remaining positions is below 1.15% of the central frequency (350 MHz) of the simulated interval.

Table 4.2: Results of the comparison between measurements and simulations with helix antennas

Slave #	SF [%]	Δf [MHz]	CTF Measurement RMS [dB]	CTF Simulation RMS [dB]
1	76%	-3.5	-27.5	-24.7
2	94%	4	-20.4	-18.3
3	94%	1.5	-19.6	-18.3
4	79%	19	-23.8	-24.5

PIFA: The similarity factor values with the PIFAs are lower than with the helix antennas. However, many similar conclusions to the case of the helix antennas can be drawn. The two central slaves have the best matching. As in the case with helix antennas, also a high position symmetry in the CTF results is clearly visualized. Additionally, most resonant frequencies are coincident between measurements and simulations. The maximum frequency offset between simulation and measurement represents 0.44% of the central frequency of the compared frequency range.

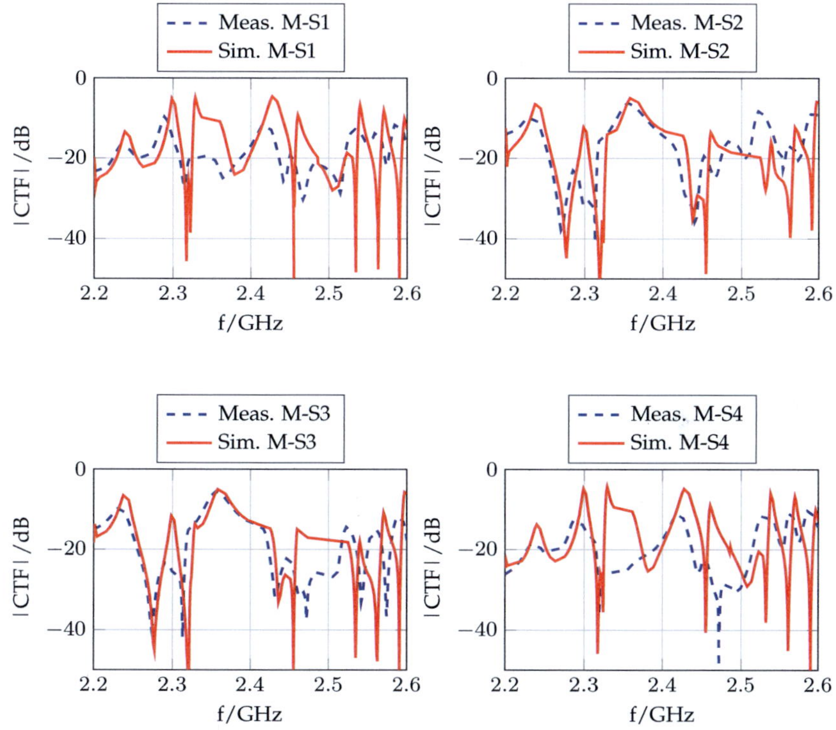

Figure 4.12: Validation of the CTF measured and simulated in the channel emulator with PIFAs

Table 4.3: Results of the comparison between measurements and simulations with PIFAs.

Slave #	SF [%]	Δf [MHz]	CTF Measurement RMS [dB]	CTF Simulation RMS [dB]
1	73%	-10.5	-16.8	-13.1
2	82%	-9	-13.1	-13.1
3	86%	-8.5	-13.9	-13.1
4	65%	-8.5	-16.6	-12.9

5 Analysis, Modeling and Characterization of the Channel

5.1 Parametrization of the Channel

In this chapter, the influence of several system parameters on the CTF is investigated: Physical dimensions of the channel emulator, amount of antennas, and position of the antennas. The study is carried out by means of parametrization. Hence, the setup is simulated many times for each particular parameter under study. In each of these simulations, the paramater under study takes on different values, while the other parameters remain constant. In this way, it can be analyzed how the variation of each individual parameter affects the wireless channel inside the emulator. Due to time and cost constrains, this parametric study was carried out only by means of electromagnetic simulations, employing the models presented in Section 4.1.

For every parameter under investigation, up to four plots of some of the obtained CTFs are shown for visual comparison. If more were plotted, the figures would be unintelligibly. Furthermore, the CTF with one particular value of the parameter under study is taken as reference case, and the remaining CTFs are compared with such a reference by means of the similarity factor.

5.1.1 Physical Dimensions of the Channel Emulator

Three physical dimensions of the channel emulator were parametrized: Length(L), width (W), and height (H). For the study of the similarity factor, the dimensions of the reference case are always the following: $W = 500\,\mathrm{mm}$, $L = 800\,\mathrm{mm}$, and $H = 50\,\mathrm{mm}$. The distance between the emulator floor and the antenna planes, i.e. the antenna height, is $Za = 29\,\mathrm{mm}$. These dimensions are also employed in the remaining simulation models of this chapter when none of the mentioned parameters is considered.

Two antennas were placed inside the battery emulator in the simulation models employed for the parametrization of the physical dimensions. One antenna was in the center of the XY-plane and the other one separated by 125 mm along the Y-axis. The S_{21} parameter is the CTF.

5.1.1.1 Length and Width

A step of 30 mm was employed for the parametrization of the length and the width, because this is a quarter of the wavelength at a frequency of 2.4 GHz, the center of the third frequency sub-interval. Fig. 5.1 and 5.2 depict some CTFs plots for the parametrization of the length and the width, respectively. The similarity factors are shown in Fig. 5.3.

The similarity factor always stays above 90% for both physical dimensions with the helix antennas at low frequencies, which proves a low

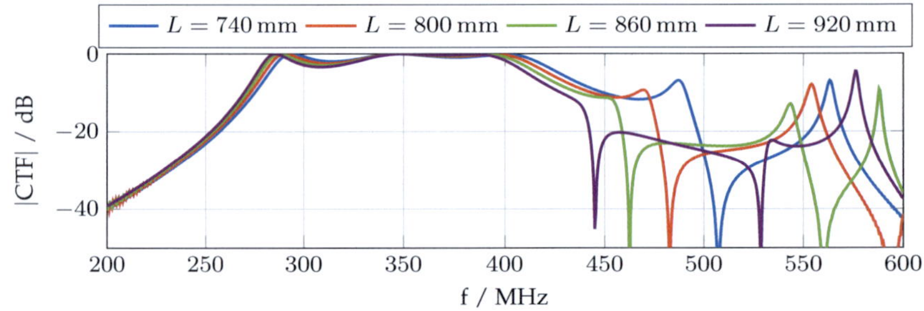

(a) Helix (LF)

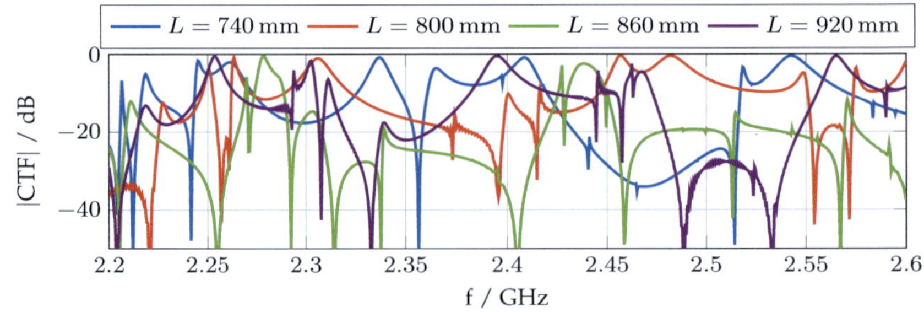

(b) PIFA (HF)

Figure 5.1: Parametrization of the battery emulator length (L)

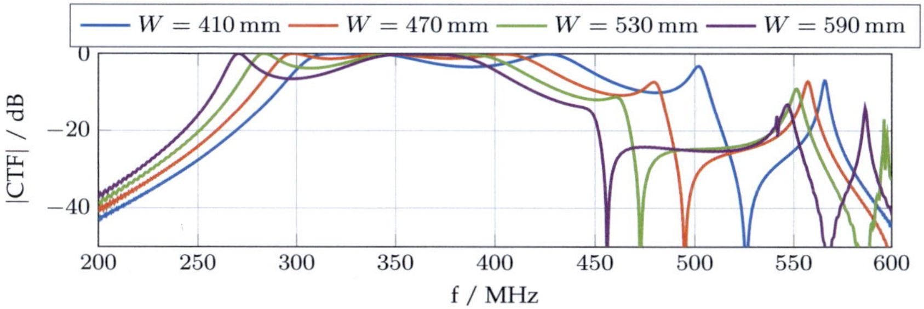

(a) Helix (LF)

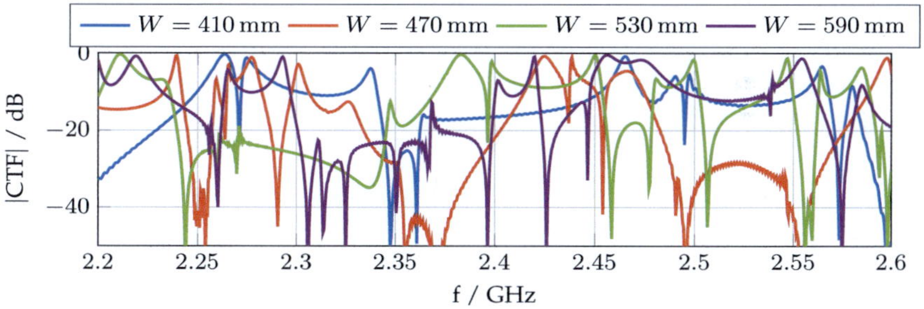

(b) PIFA (HF)

Figure 5.2: Parametrization of the battery emulator width (W)

influence of these dimensions on the CTF in the studied cases. Similar conclusions can be drawn by observing the CTF plots in Fig. 5.1(a) and Fig. 5.2(a): All the plotted CTFs are very similar in their forms. However, they differ in the bandwidth of the frequency range in which the CTF presents an almost constant gain. This can be visualized with the position of the first strong notch of the CTF, e.g. the one at a frequency of 483 MHz in the reference case. When the length or the width increases, the notch frequency is smaller, and the bandwidth of the almost constant gain region decreases.

On the other side, the variation of the length and the width of the battery emulator clearly plays a very important role for the CTF at higher frequencies using the PIFAs. The similarity factors are always considerably smaller than at lower frequencies. As already explained in Section 3.4.5, the bigger the emulator dimensions, the bigger the density of resonant modes is, due to the resonant behavior of the channel emulator. This increases the frequency-selectivity of the channel. The CTF plots can be observed in Fig. 5.1(b) and Fig. 5.2(b) for the parametrization of the length and the width, respectively.

5.1.1.2 Height

For the parametrization of the height, a smaller step was employed: 5 mm. At low frequencies, the similarity factor also keeps above 90% for the simulated values of height, as it is shown in Fig. 5.4(c). Looking at the CTF plots in Fig. 5.4(a), the lowest frequency limit of the almost constant gain region increases when the height becomes bigger, reducing the bandwidth of the mentioned region. The attenuation along this almost constant gain region, as well as the frequency position of the notch located at 483 MHz in the reference case, remain practically constant in the simulated cases.

In the higher frequency range, the similarity factors are relatively high for the simulated values of height up to 55 mm. Observing Fig. 5.4(b), the CTFs for these height values have the same resonant frequencies, although the bandwidth of the CTF around these frequency values decreases when the height increases. On the other hand, the situation is completely different for the highest values of height: The similarity factor decreases much faster when the height increases, as well as the number of resonant frequencies in the CTF.

The last phenomenon can be understood following the explanation in Section 3.7.1. There is a certain frequency value, $f_{\min(l>0)}$, from which resonant modes with the index $l > 0$ appear. According to (3.4), when

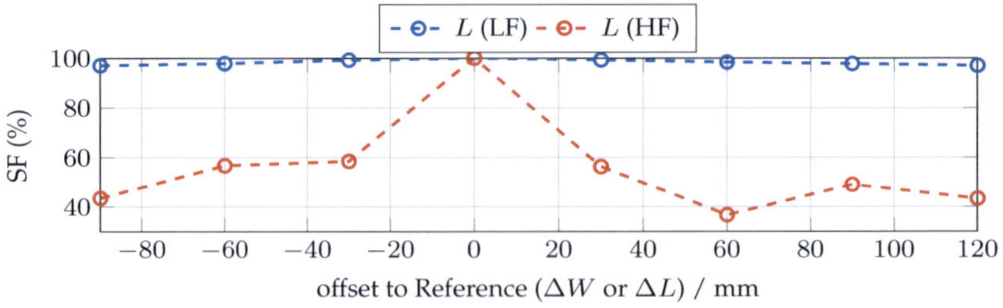

Figure 5.3: Similarity factor of the parametrization of W and L

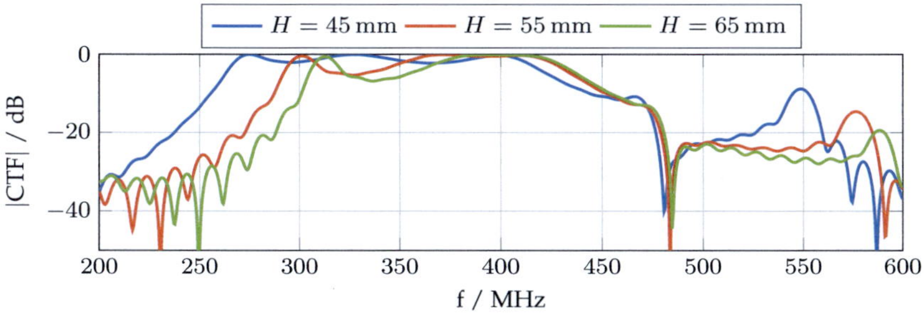

(a) Variation of the CTF with H at low frequencies

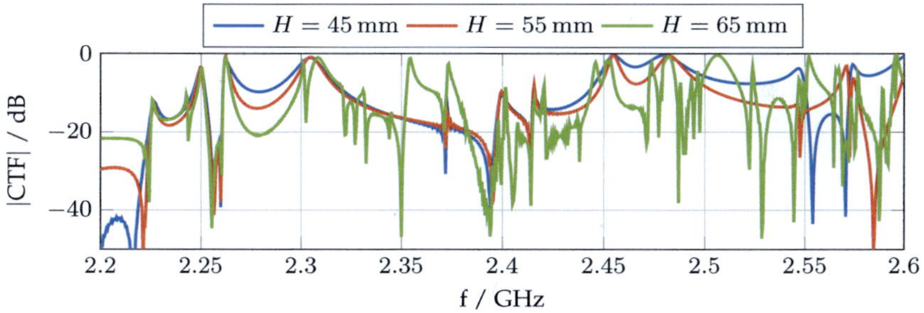

(b) Variation of the CTF with H at high frequencies

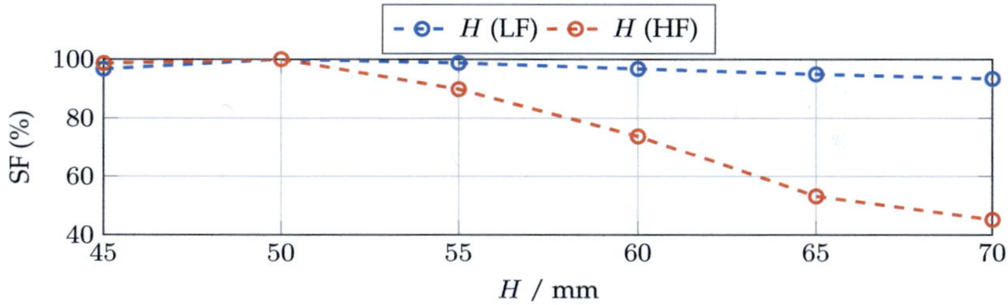

(c) Variation of the similarity factor with H

Figure 5.4: Parametrization of the battery emulator height (H)

$l = 0$ the values and amount of resonant frequencies do not depend on the height. Therefore, the height does not influence the CTF significantly. Table 5.1 depicts the values of $f_{\min(l>0)}$ calculated for each simulated height[1]. $f_{\min(l>0)}$ is above the simulated frequency range for the cases with $H \leq 55\,\mathrm{mm}$, i.e. it is higher than 2.6 GHz. This is the reason why the height does not have a big influence on the CTF for these height values. The case $f_{\min(l>0)} = 2.6\,\mathrm{GHz}$ occurs when $H = 55.7\,\mathrm{mm}$. Therefore, for the remaining height values, which are bigger than 55.7 mm, the emulator height plays an important role in the amount and values of the resonant frequencies. Hence, the height has a greater impact on the CTF and the similarity factors decrease. For these height values, as expected, the amount of resonant frequencies increases with the height.

Table 5.1: $f_{\min(l>0)}$ as a function of the simulated emulator height

Height [mm]	45	50	55	60	65	70
$f_{\min(l>0)}$ [GHz]	3.34	3.01	2.73	2.51	2.32	2.15

5.1.2 Amount and Position of the Slaves

Firstly, the same reference case as in Section 5.1.1 was employed but with the distance between the antennas as parameter under investigation. The position of the antenna in the center of the emulator remains constant, but the position of the second one varies along the Y-axis, parallel to the emulator length. The distance between the antennas varies from 125 mm to 335 mm, with a step of 35 mm.

A low influence of the antenna position on the CTF at low frequencies can be observed in the plots in Fig. 5.5(a). This is also verified by the high similarity factor values shown in Fig. 5.5(c). However, the distance between the antennas has an important influence on the channel attenuation: The bigger the distance, the higher the attenuation. The bandwidth of the almost constant gain region also decreases when the distance increases.

The channels present a higher position dependence of the CTF form at higher frequencies, what is proved by the lower similarity factor values. Nevertheless, the values of most resonant and notch frequencies are very similar in every simulated case. Additionally, there is not a clear

[1] $f_{\min(l>0)}$ is calculated for an ideal rectangular cavity resonator only as reference value.

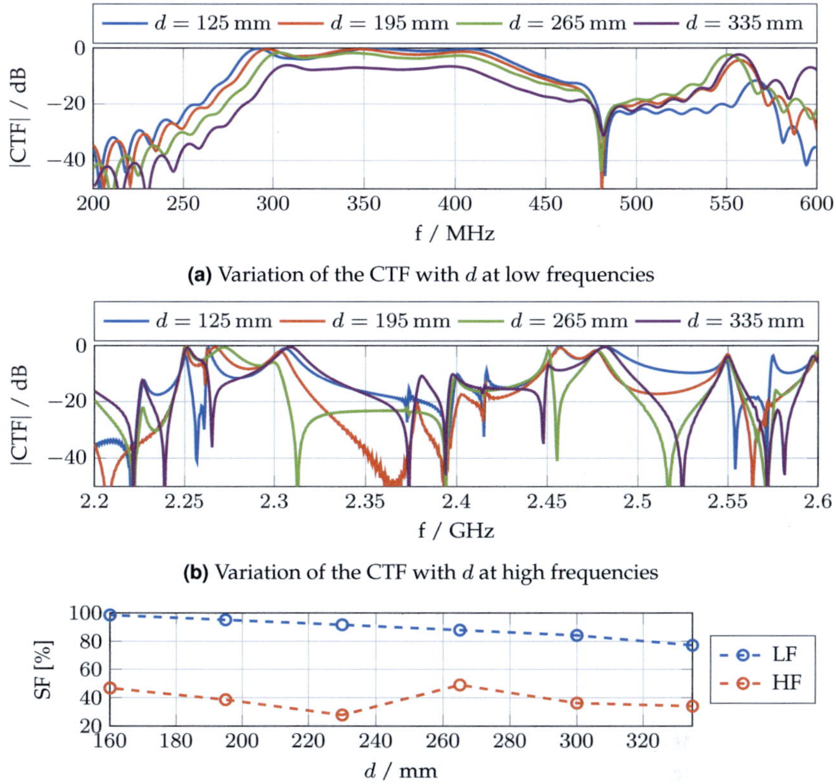

(a) Variation of the CTF with d at low frequencies

(b) Variation of the CTF with d at high frequencies

(c) Variation of the *SF* with d

Figure 5.5: Parametrization of the distance between master and slave (d)

relationship between the channel attenuation and bandwidth around the resonant frequencies, as in the case of lower frequencies.

Four setups with different amounts of slaves were simulated for each antenna type but always using the small channel emulator, in order to expand the evaluation of the impact of antenna positions on the channel properties, as well as the influence of the amount of antennas inside the emulator. Every setup contains one master antenna but one of the following amounts of slaves: 4, 8, 16, and 30. Each of these configurations are called "architectures" and their corresponding denotations are

1M-4S, 1M-8S, 1M-16S, and 1M-30S, respectively. The first architecture was already used for the validation of the simulation models against the measurement results. Some of the remaining modeled architectures are shown in Fig. 5.6. As previously, only one figure with four of the obtaining CTF curves is presented for each architecture and antenna type. Otherwise, the figures would be incomprehensive. However, the plotted CTFs were selected strategically, in order to avoid the cases of highest symmetry.

The results of CTFs for the four architectures with helix antennas at low frequencies are shown in Fig. 5.8. For the cases with four and eight slaves, the high similarity in form between the different positions among one particular architecture is clearly visible. The low position dependence of the CTF for the case 1M-4S with helix antennas has been already commented and the architecture 1M-8S follows the same concept. The bandwidth of the almost constant gain region is a bit shorter with eight than with four slaves. However, this low position dependence characteristic of the CTF is not clearly observable in the architecture 1M-16S and especially in 1M-30S. The results in Fig. 5.7(a) also prove how the similarity factor values are smaller for the two architectures with

(a) Helix, 1M-8S **(b)** Helix, 1M-16S **(c)** PIFA, 1M-30S

Figure 5.6: Simulation models of some of the combinations of architecture and antennas type inside the battery emulator

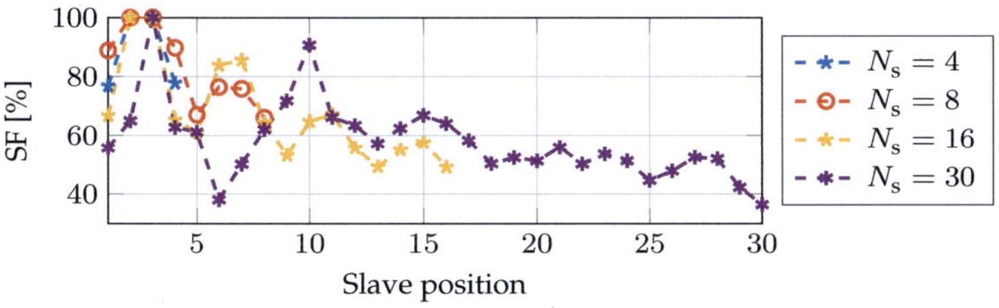

(a) Helix antennas at low frequencies

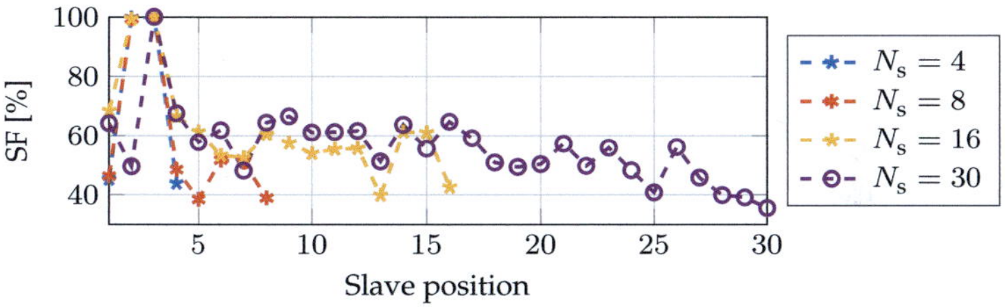

(b) PIFAs at high frequencies

Figure 5.7: Variation of the *SF* with the slave position (with N_s slaves inside the battery emulator)

a bigger amount of antennas. A reason for this is that the greater the number of antennas, the smaller is the separation between them. This increases the mutual coupling between the antennas. In all the simulated architectures, the positions of the antennas play again an important role in the attenuation of the CTF: The bigger the distance between slave and master, the higher the attenuation. All the conclusions drawn below are probed again in Section 5.5.3 by means of the channel characterization.

When the scenario 1M-4S with PIFAs at higher frequencies was previously analyzed in other Sections of this work, it was concluded that the CTF was strongly position-dependent. This is a characteristic of the four simulated architectures and, therefore, most of the similarity factor results in Fig 5.7(b) are very small. Many resonance frequencies of the CTFs at different positions are practically coincident in the architectures 1M-4S and 1M-8S m, similar to the case of the parametrization of the distance between antennas. However, this is not the case for the remaining architectures with more antennas. The higher mutual coupling plays again an important role.

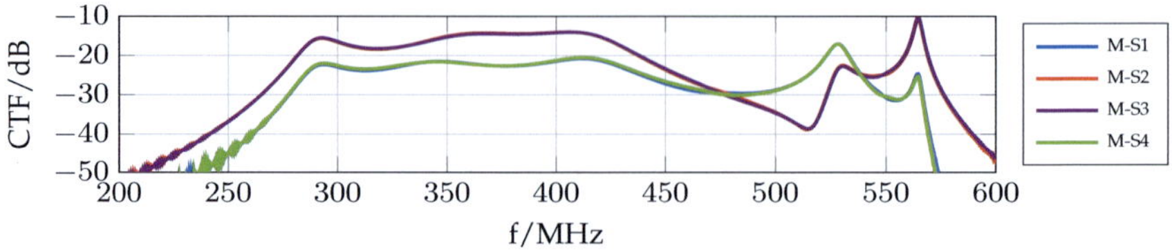

(a) 1M-4S

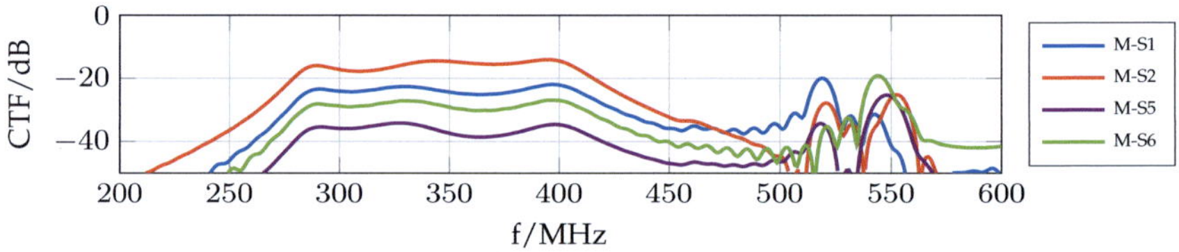

(b) 1M-8S

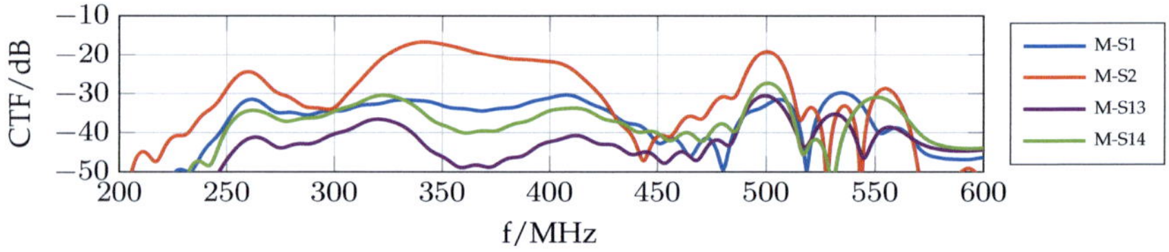

(c) 1M-16S

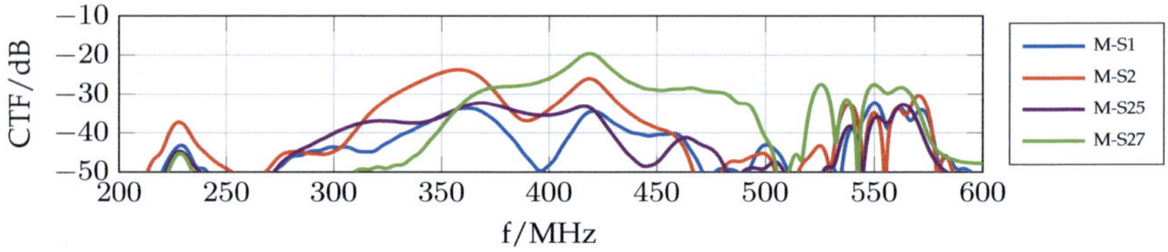

(d) 1M-30S

Figure 5.8: CTFs for different architectures at low frequencies

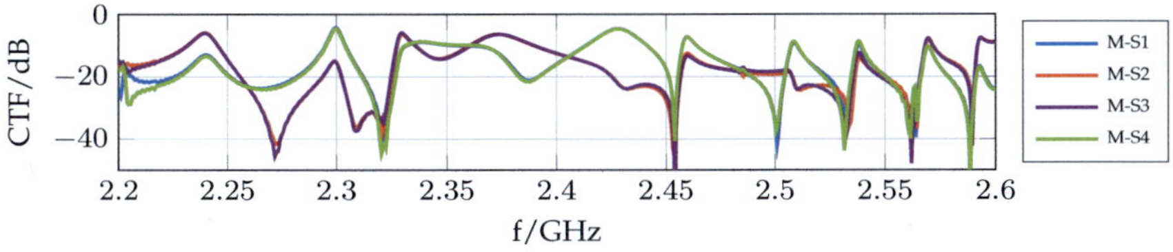

(a) 1M-4S

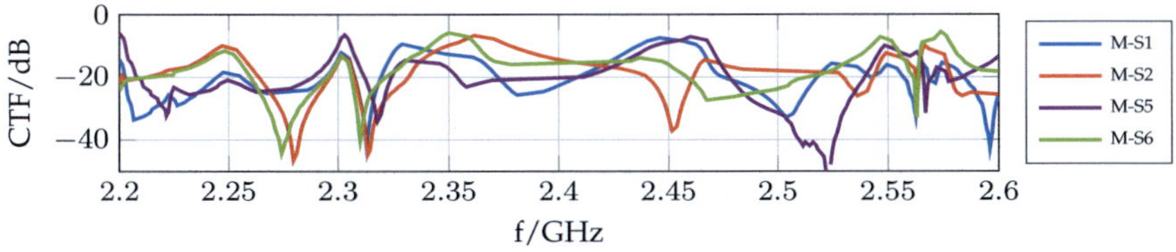

(b) 1M-8S

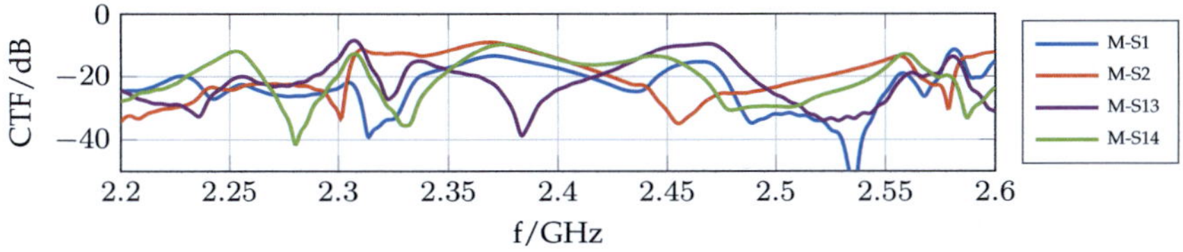

(c) 1M-16S

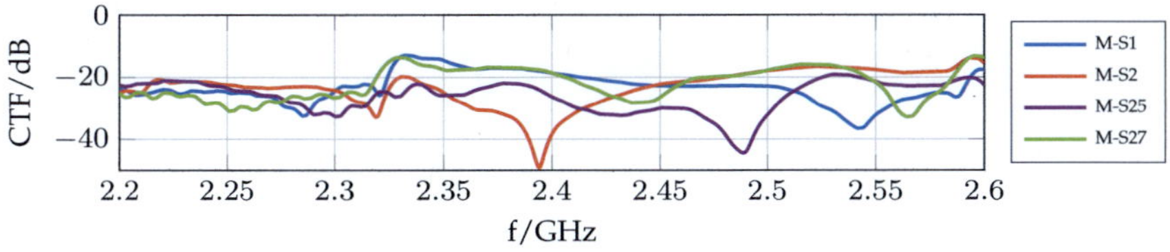

(d) 1M-30S

Figure 5.9: CTFs for different architectures at high frequencies

5.1.2.1 Height of the Antennas (Z-Axis)

The reference case in this test is again the same one as in Section 5.1.1. The parameter under investigation is the height of the antennas inside the emulator (Z_a), i.e. the distance between the antenna plane and the emulator floor.

In this parametrization, the same conclusions can be drawn for both frequency ranges looking at the results in Fig 5.10. The attenuation at the resonant frequencies remains practically constant with the variation of the antenna height, but the smaller the antenna height, the smaller the coherence bandwidth around the resonant frequencies. Near the metallic walls, the electric and magnetic fields are smaller due to the boundary conditions and, hence, the CTF decreases faster outside the the resonant frequency region. The similarity factors are very similar for

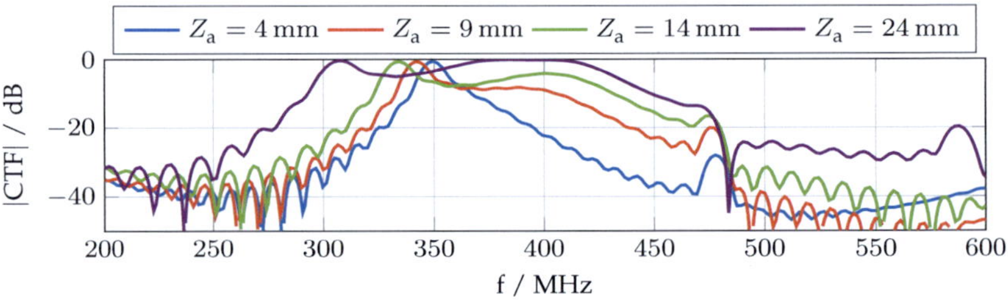

(a) Variation of the CTF with Z_a at low frequencies

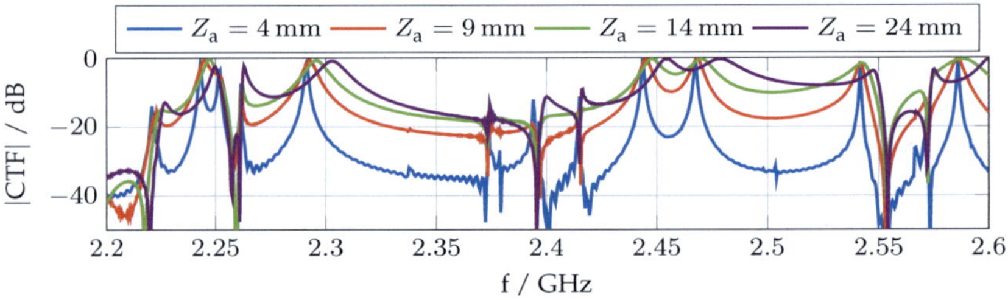

(b) Variation of the CTF with Z_a at high frequencies

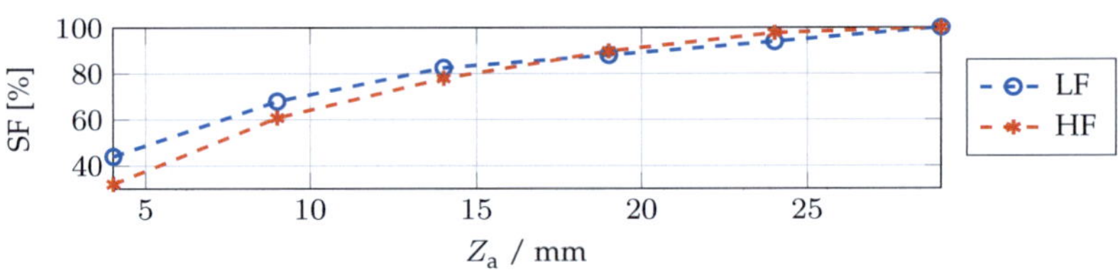

(c) Variation of the *SF* with Z_a

Figure 5.10: Parametrization of the height of the antennas (Z_a, antenna position in Z-axis) both frequency intervals. The smaller the antenna height, the smaller the similarity factor.

5.2 Channel Modeling for Simulations

As most wireless communication systems, the wireless approach for the in-battery communication operates with bandpass signals. The spectrum of this type of signals occupies only a certain frequency range and it is centered around a frequency value called center frequency f_c (also named carrier frequency). If the signal bandwidth is B, then the frequency range of the channel occupied by the bandpass signal is $\left[-\frac{B}{2} + f_c, f_c + \frac{B}{2}\right]$ [72]. In the case of the wireless approach for the BMS, the medium is not shared with external wireless devices, due to the shielding provided by the metallic housing. However, working at frequencies much higher that the required bandwidth is mandatory, in order to employ antennas with a reasonable size. Therefore, also in this work, the proposed communication system will operate with bandpass signals.

Most digital communication systems first generate a baseband signal (i.e., a signal with $f_c = 0$) at the transmitter. Then, this signal is converted into a bandpass signal by shifting the baseband signal to a higher frequency interval, centering the baseband signal around the desired center frequency. In the receiver, the inverse process is carried out.

The channel impulse response (CIR) is a characterization of the channel that contains the complete required information for the analysis of the transmission of a signal though this channel [74]. For software-based simulations of the complete communication system, the impulse response of a multipath channel can be modeled as a finite impulse response (FIR) filter. Typically, wireless channels are time-variant, due to the relative motion between transmitter, receiver, and/or the environment. In such cases, the filter coefficients must also be time-variant. However, there is no relative motion between antennas, or between an antenna and the environment inside the battery. Furthermore, the metallic housing provides an electromagnetic shielding against any reflective object located outside the battery. Consequently, the wireless in-battery channel can be considered as practically static, i.e. time-invariant [104]. Such a channel can be modeled as a time-invariant FIR filter.

Due to the importance of the CIR measurement, several channel sounding techniques exist [61, 74]. In this work, the technique employed was a frequency domain channel sounder, using a VNA[2]. With this measurement technique, the CIR is obtained in the frequency domain, i.e. the CTF. As the channel is assumed to be time-invariant, and the measured CTF is represented discretely, the CIR response is obtained by converting the CTF to the time domain with the inverse discrete Fourier transform (IDFT) or, as in most software tools, with the inverse fast Fourier transform (IFFT).

Only the frequency range of the CTF that corresponds to the bandpass signal is of interest. Therefore, both this range of the CTF and the bandpass signal can be shifted to the frequency range $\left[-\frac{B}{2}, +\frac{B}{2}\right]$, centered around the null frequency. This new representation of the CTF is called complex baseband equivalent model of the channel. The shifted bandpass signal is now a baseband (also called lowpass) signal, which is the complex envelope of the bandpass signal [74]. For a software-based simulation of a communication system, working with baseband signals and the basedband equivalent model of the channel permits the use of a smaller sampling frequency, i.e. larger time steps, than the ones required to work with bandpass signals. This boosts the simulation efficiency and decreases the simulation duration [14, 98].

The relationship between the communication system bandwidth and the sampling frequency must be remarked. In representations of real signals, the frequency spectrum is guaranteed to be Hermitian, i.e. it always follows $X_{(\omega)} = X^*_{(-\omega)}$. Therefore, due to this spectrum redundancy, the bandwidth of the signal is measured from the null frequency up to the maximum frequency. Following the Nyquist theorem, this maximum frequency must be lower than half of the the sampling frequency f_s, i.e. $B < \frac{f_\mathrm{s}}{2}$. In the case of complex signals, the quadrature sampling is used. This outputs two separate sequences of samples, one representing the in-phase component, and another representing the quadrature component. The complex signal is generated from these two sequences. Due to the complex nature of this quadrature signal, its spectrum is not forced to present a symmetry with respect to the 0 Hz axis [51]. Then, the bandwith

[2] Other channel sounding techniques have been implemented in the framework of a master thesis [114]. However, the VNA offered many advantages over the software defined radio-based alternatives for the application in this work.

of complex signals is defined as the difference between the maximum positive and minimum negative frequency. Hence, for complex signals the sampling frequency must be only greater than the signal bandwidth, i.e. $B < f_s$, but not twice as for real signals.

5.3 Modeling of the Channel Noise

In the framework of the IntLiIon project, the noise in a hot battery was measured only up to a frequency of 100 MHz, which is below the frequency range of interest in this work. No reference in literature or published papers has been found with this kind of measurements at higher frequencies. Therefore, the noise in the wireless channel within the battery at the frequencies of interest was modeled based on the analysis of the three possible noise sources for the in-battery channel:

1. Thermal noise
2. The power train components (e.g. inverter)
3. External noise or interference sources (e.g. other wireless services operating in the same frequency range)

However, the last two noise sources can be neglected. The power train components produce high noise power only in frequency ranges below the ones considered in this work [16]. And regarding external noise or interference, they are prevented by the shielding provided by the metallic housing. The remaining noise source, the thermal noise, is modeled as additive white Gaussian noise (AWGN). Therefore, the noise in the wireless in-battery channel is also modeled as AWGN in this work.

The noise power density (N_0) can be calculated as

$$N_0 = kT, \tag{5.1}$$

where k is the Boltzmann's constant in Joules per Kelvin, and T is the temperature in Kelvin of the system[3]. According to [73], typical operation

[3] (5.1) is actually only valid when the antenna and the receiver input are perfectly matched, which is not the case in the simulation or measurement results. However, as later explained in Section 5.4, the system is intended to operate in frequency intervals around resonant frequencies, where there is typically a good matching between antenna and receiver input.

temperature range of Li-ion batteries for vehicles is between -40°C to $+50^{\circ}$C, which would lead to a maximum $N_0 = -173.5\,\mathrm{dBm/Hz}$. This value will be used as noise power density of the channel in the remaining work.

5.4 Selection of the Center Frequency

Every communication node (master and slaves) in the communication network employs the same center frequency. The selection of the center frequency plays a key role in the performance of the communication system, since its variation can have a great influence on the passband channels between master and each slave.

The system should ideally operate in the frequency range with the best signal-to-noise ratio (SNR) between the master and every slave. However, this is not possible because the channels between the master and each slave are different. A trivial solution for the selection of the center frequency might be to look for the frequency with the highest average SNR among all the slaves. But at this center frequency one or several slaves might present a very inconvenient passband channel. For example, a notch frequency very close to the center frequency. The BMS cannot manage the battery safely if some of the slaves cannot communicate properly. Therefore, this solution would be sub-optimal. A better algorithm for the selection of the center frequency had to be developed, with the target of ensuring a good compromise of SNR among all the slaves [104].

5.4.1 Algorithm for the Center Frequency Selection

The SNR at different receivers and at different frequencies must be compared to look for an appropriate center frequency. However, due to the assumption of AWGN, the noise power spectral density is considered constant along the entire frequency range and at every receiver. Then, comparing the received signal power is equivalent to compare the SNR.

The first step of the algorithm is to calculate the received signal power at every slave and for each center frequency candidate value. The transmission power is normalized to 1 W for simplicity. The "complete" baseband equivalent channel must be obtained for each slave and for each center frequency candidate value to achieve this. "Complete" means that not only the in-battery channel must be considered, but also the transmit and receive filters employed at the transmitter and receiver, respectively[4]. Therefore, the transmit and receive filter coefficients[5] , as well as the sampling frequency (which determines the communication system bandwidth) must be known in advance to employ this algorithm.

The frequency range under analysis, i.e. the first frequency sub-range for the helix antennas and the third one for the PIFAs, is divided into a set of equidistant frequency values, which are the center frequency candidate values. Being i the index of the frequency values and l the index of the slaves (or equivalently, of each CTF), the following steps are performed for every candidate frequency $f_{(i)}$ and for every slave:

- The CTFs of the baseband equivalent channels between the master and each slave are calculated (in discrete frequency domain), employing the different frequencies $f_{(i)}$ as center frequency. Then, these CTFs are interpolated to have the same length K as the transmit and receive filter impulse responses. The result obtained for every i and l values is $H_{\mathrm{ch}(i,l)}[k]$, where k represents the discrete frequency domain.
- The received power at the output of the receive filter for each $((i,l))$-case is calculated in the discrete frequency domain as:

$$P_{(i,l)} = \frac{\sum_{k=0}^{k=K-1} \left| H_{\mathrm{TxF}}[k] \cdot H_{\mathrm{ch}(i,l)}[k] \cdot H_{\mathrm{RxF}}[k] \right|^2}{K^2} \tag{5.2}$$

  where $H_{\mathrm{TxF}}[k]$ and $H_{\mathrm{RxF}}[k]$ are the impulse responses in frequency domain of the transmit and receive filters, respectively; and $P_{(i,l)}$ is the received power at the slave l when $f_c = f_{(i)}$.

Next, a new vector is created, $P_{\mathrm{min}(i)}$, with the same length I as the vector of the frequencies $f_{(i)}$. In each position i the minimum received power at

---

[4] In [104], the complete channel was assumed as the in-battery channel plus the transmit filter. In this work, the receive filter is added, since it also modifies the E_b/N_0 observed by the demodulator at the receiver.

[5] The components of the communication system are explained in detail in Section 6.1.

the corresponding $f_{(i)}$ among the slaves is stored[6].

$$P_{\min(i)} = \left[\min\left(P_{(1,l)}\right), \min\left(P_{(2,l)}\right), \cdots, \min\left(P_{(I,l)}\right)\right] \quad (5.3)$$

The frequency value $f_{(i)}$ corresponding to the maximum value among the vector $P_{\mathrm{min(i)}}$ is selected as the system's center frequency. The center frequency value obtained by means of this algorithm is referred to as *optimal center frequency* in the remaining work.

5.4.2 Results

The algorithm has been employed to select the optimal center frequency for each of the four architectures simulated in Section 5.1.2, with both evaluated antennas and frequency sub-ranges. However, as mentioned in the description of the algorithm, some communication system parameters must have been previously defined. The communication system configuration described in Section 6.1 was used, which outputs a QPSK-modulated signal with a bit rate of 2 Mbps. This configuration employs matched root-raised-cosine-filters with a roll-off factor of 0.2 at both transmitter and receiver, and a sampling frequency of 4 MHz. The obtained center frequency values are presented in Table 5.2.

Table 5.2: Optimal f_c values calculated for scenarios simulated in Section 5.1.2

Antenna type	f_c [MHz]			
	1M-4S	1M-8S	1M-16S	1M-30S
PIFA	2334.5	2338.5	2557	2513.7
Helix	412	326	320.5	416.4

6 If at this point the average received power instead of the minimum would have been stored, it could happen that for this frequency some slaves present a good CTF but one of them a very bad one, and then the performance of the entire system would be degraded.

5.5 Channel Charaterization

5.5.1 Characterization of the In-Battery Channel

In a similar way to most real wireless channels, as for example in mobile communications, a signal that travels from a transmitter to a receiver inside an automotive battery is exposed to multipath fading. That means that the transmitted signal is reflected multiple times and it reaches the receiver though multiple paths. Inside an automotive battery it can be reflected by the metallic housing, the cell cans, etc. A channel that produces this impact on the transmitted signal is known as a fading channel [82]. Such a channel modifies the amplitude, phase, and angle of arrival of the received signal.

There are two different types of fading: large-scale and small-scale. Large-scale fading refers to the variation of the signal due to its propagation over large distances, i.e. much bigger than the wavelength of the signal. The large-scale fading is typically dominated by the terrain characteristics. This is, however, not of interest in the case of wireless in-battery communications, due to the relationship between the dimensions of the environment and the investigated wavelength range. On the other hand, small-scale fading represents the strong signal variations along short displacements, even shorter than the wavelength of the signal. This type of fading is the one that must be analyzed for the in-battery communication.

The small-scale fading alters the received signal by two ways. The first one is producing a time-spreading of the signal, due to the different times of arrival of the multipath components. The second one is due to the time-variant behavior the channel [82]. The last one, however, can be practically ignored because, as already explained in Section 5.2, the in-battery channel can be assumed to be time-invariant. Therefore, only the time-spreading characteristics of the channel must be analyzed. Two parameters are employed for the characterization of the time-spreading behavior of a channel: the delay spread and the coherence bandwidth.

Delay Spread: The delay spread is analyzed in the time-domain and it indicates the dispersion of the power delay profile (PDP) over the time. The PDP is obtained as $|h(\tau)|^2$, where h is the CIR and τ the time over which the h is spread [74, 82]. The higher the multipath characteristic of the channel, the higher the delay spread. This can be characterized by two metrics: The mean delay spread ($\bar{\tau}$) and the RMS delay spread (σ_τ). They can be calculated with the following equations[7]:

$$\bar{\tau} = \frac{\int_0^\infty \tau \left|h(\tau)\right|^2 d\tau}{\int_0^\infty \left|h(\tau)\right|^2 d\tau} \tag{5.4}$$

$$\sigma_\tau = \sqrt{\frac{\int_0^\infty (\tau - \bar{\tau})^2 \left|h(\tau)\right|^2 d\tau}{\int_0^\infty \left|h(\tau)\right|^2 d\tau}} \tag{5.5}$$

.

Coherence Bandwidth: The coherence bandwidth B_{coh} is obtained in the frequency domain, i.e. from the CTF. It is a metric of the frequency range over which the channel exhibits an approximately constant gain and a linear phase. That means, that every frequency component in this frequency range is affected by the channel in a similar way. Therefore, the amplitude of the frequency components are strongly correlated [29, 74, 82]. The coherence bandwidth is mathematically defined as the bandwidth over which the normalized auto-correlation function of the CTF remains above a pre-defined threshold. Different values have been given to this threshold in literature, for example 0.9 [41] or 0.5 [1].

The RMS delay spread and the coherence bandwidth are inversely proportional, i.e.

$$B_{\text{coh}} = \frac{\alpha}{\sigma_\tau}, \tag{5.6}$$

where α is a constant, whose value depends on the channel multipath structure and the threshold value employed for the calculation of the coherence bandwidth [29, 74, 82]. There are, however, several published results for this relationship between RMS delay spread and the coherence bandwidth under typical wireless channels. For example, a value of $\alpha = 0.276$ was reported for an urban environment [1], and $\alpha = \frac{1}{2\pi}$ for the ionosphere [6].

[7] If the signals under analysis are discrete, the integral operation in (5.4) and (5.5) is replaced by a summation.

The fading characteristic of a channel can be classified as "flat" or "frequency-selective". In the first case, it can be assumed that the channel affects every signal frequency component in a similar manner. On the other hand, a channel is referred to as frequency-selective when it induces inter-symbol interference (ISI) to a signal that is transmitted though this channel. ISI means that the duration of a transmitted symbol is considerably extended beyond the beginning of the next transmitted symbol, degrading the quality of the transmitted signal. This issue can be resolved at the receiver by the employment of techniques such as channel equalization, but this increases the receiver complexity [72, 83].

The fading characteristic of a channel and, consequently, the occurrence of ISI, depends not only on the channel but also on the bandwidth of the transmitted signal. A channel can be considered to be flat and not to induce ISI under the following conditions [74, 82]:

$$T_{\text{sym}} > \sigma_{\mathcal{T}} \tag{5.7}$$

and

$$B_{\text{sym}} < B_{\text{coh}}, \tag{5.8}$$

where T_{sym} is the symbol duration and B_{sym} is the symbol bandwidth. The design of the communication system is discussed in Chapter 6. There, only the performance of phase-shift keying (PSK) modulations is analyzed. In this type of modulation, the relationship between symbol duration and bandwidth at the output of the modulator is $B_{\text{sym}} = \frac{1}{T_{\text{sym}}}$. However, due to the employment of a transmit filter, for example a root-raised-cosine filter, the symbol is spread and hence, its bandwidth is enlarged [72]. Consequently, it can be said that

$$B_{\text{sym}} \geq \frac{1}{T_{\text{sym}}}. \tag{5.9}$$

Then, if (5.8) is accomplished, the following happens with the relationship between T_{sym} and $\sigma_{\mathcal{T}}$, employing (5.6), (5.8), and (5.9) together:

$$\frac{1}{T_{\text{sym}}} \leq B_{\text{sym}} < B_{\text{coh}} = \frac{\alpha}{\sigma_{\mathcal{T}}} \tag{5.10}$$

and then,

$$\frac{\sigma_{\mathcal{T}}}{T_{\text{sym}}} \leq \alpha. \tag{5.11}$$

As α is always smaller than one in literature, then (5.7) is accomplished though (5.11). That means, that only by calculating the coherence bandwidth of the channel and comparing it to the transmitted signal bandwidth, it can be determined whether the channel presents frequency-selective or flat fading.

5.5.2 Method

The calculation of the coherence bandwidth was carried out for every slave in each of the four simulated architectures (with 4, 8, 16, and 30 slaves) and with both antenna types. The coherence bandwidth was calculated for the entire simulated spectrum in every case, i.e. between 200 MHz and 600 MHz for the helix antennas at lower frequencies (as shown in Fig. 5.8), and between 2200 MHz and 2600 MHz for the PIFAs (as depicted in Fig. 5.9). The threshold value for the calculation of the coherence bandwidth was set to 0.5. With this value, the coherence bandwidth of an ideal channel with constant gain and linear phase is equal to the sampling frequency, i.e. to the frequency range of the channel under analysis.

5.5.3 Results and Discussion

The results of the coherence bandwidth are shown in Table 5.3. These results support the conclusions drawn in Section 5.1.2. With four, eight, and 16 slaves, the coherence bandwidth is higher at lower frequencies. That means that the frequency range where the channel presents flat fading is larger. The frequency range with flat fading is shorter at low frequencies only with the largest number of slaves (30). This characteristic of the channel at low frequencies and with big numbers of slaves was already discussed in Section 5.1.2. Additionally, the standard deviation of the coherence bandwidth is much smaller at lower frequencies, except for the case with eight slaves. This supports the conclusion of lower position dependence at lower frequencies for reduced amounts of slaves.

As already explained in Section 5.1.1.1, the length and width of the emulator do not have a significant influence on the CTF's form for a reduced number of slaves and at low frequencies, as long as both length and width remain in the same order of magnitude as the wavelength. This fact produces a interesting synergy with the large coherence bandwidth of the

Table 5.3: Results of the coherence bandwidth obtained for the simulated channels with both antenna types

Arquitecture	B_{coh} [MHz]			
	Helix		PIFA	
	Min	Mean $\pm$ Std	Min	Mean $\pm$ Std
1M-4S	197.9	233.6 ± 40.7	91.2	183.0 ± 105.1
1M-8S	86.2	149.4 ± 54.9	73.9	101.8 ± 23.6
1M-16S	142.8	228.7 ± 46.4	97.9	214.9 ± 82.0
1M-30S	88.8	161.1 ± 45.9	132.17	261.2 ± 70.9

channels: Since the large frequency range of practically flat attenuation is hardly influenced by the antenna position, the center frequency could be chosen arbitrary along this frequency range, without requiring the use of the algorithm from Section 5.4. Consequently, it would not be necessary to measure the CTF between the master and every slave, as the mentioned algorithm demands. Alternatively, measuring the CTF between one master and only one slave would be enough, since the remaining ones can be assumed to be very similar and present an almost coincident frequency interval with almost constant gain and high coherence bandwidth. This would be a simple *plug-and-play* system, i.e. a system that can be used without modifications or previous calibration in different batteries, as long as the mentioned conditions are met. However, this *plug-and-play* concept is not considered in this work.

6 Design and Simulation of the Physical and the MAC Layer

Based on the analysis of the channel, a proper communication system must be designed, capable of fulfilling the project's requirements on the physical and the MAC layer. In this chapter, the physical layer of a transceiver (transmitter and a receiver) is designed and simulated by means of Matlab and Simulink. The performance of this transceiver is evaluated under channels obtained in the last chapter.

The MAC layer is described at the end of this chapter. Its implementation in software, as well as its prototyping of both physical and MAC layers, is introduced in the next chapter.

Several channel coding techniques have been evaluated in [112] with the physical layer designed in this chapter and under channels also considered in this work. Since the topic of channel coding is not the focus of this work and no new channel coding technique has been developed, channel coding is not included in this chapter. Nevertheless, the selected channel coding technique in [112] for this wireless in-battery communication system was implemented in rapid prototype platforms. These results are presented in Chapter 7.

6.1 Design of the Physical Layer

The design of the transmitter involves the architecture of transmitter and receiver (modulation, synchronization, etc.), as well as the calculation of the required power for the transmitted signal. The issues of the channel that have an influence on the design of the physical layer have been already clarified in Chapter 5. Two characteristics of the channel must be particularly taken into account:

- Attenuation, which has a great influence on the received signal power
- Multipath fading characteristic, for the architecture of the transceiver

Regarding the project requirements on the physical layer, a minimum BER of 10^{-3} without channel coding has been specified at the beginning of the IntLiIon project[104]. However, there was no definition about the transmission power and transceiver architecture. The transmission power must be kept as low as possible, in order to reduce power consumption. The transceiver architecture must be as simple as possible, in order to reduce the cost and the size of the future implementation. However, the transmission power and the design of the transceiver are not independent concepts, since the BER is related to both of them. A more sophisticated transceiver may achieve a better performance, so it can reach a particular BER with lower transmission power than a simpler transceiver. Additionally, the bandpass channel between two communication devices depends on the central frequency of operation and the bandwidth of the transceiver, as explained in Section 5.2.

The entire communication modem is expected to be very small to fit into the size of a small automotive battery cell. Therefore, a digital transceiver is considered, where most of the signal processing is carried out digitally in the baseband. The particular case of a direct-conversion transceiver (also called zero-intermediate frequency transceiver) [60] was taken into account.

Fig. 6.1 highlights a scheme of a typical digital transceiver [115, 96]. From a very general and abstract point of view, the signal generation in the transmitter involves the following steps [36, 63, 96]

- The signal to be transmitted is digitally generated in the baseband
- Conversion from digital to analog is performed by means of an I/Q (dual channel) digital to analog converter (DAC)
- Filtering of the signal to avoid spectral repetitions, due to the signal sampling
- Up-conversion of the signal, i.e. shifting it to the central frequency
- The signal is filtered, to avoid interfering with the spectrum outside the transmitter bandwidth, and amplified

The steps are the opposites and in reversed order in the receiver:

- Low noise amplification of the signal

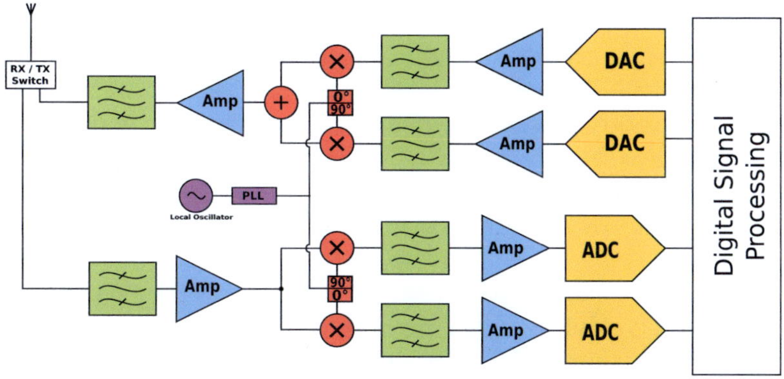

Figure 6.1: Block diagram of the digital transceiver [115]

- Down-conversion of the signal, i.e. shifting it to the zero-frequency
- Filtering of the signal to avoid aliasing
- Conversion from analog to digital is performed by means of an I/Q (dual channel) analog to digital converter (ADC)
- Digital processing of the signal in the baseband (synchronization, demodulation, decoding, etc.)

All the simulations of the communication system in this chapter correspond only to the digital processing of the signal in baseband. Therefore, in order to simulate the impact of the channel on the signal, the baseband equivalent channel models are employed. The generation of these models was already explained in Section 5.2.

6.1.1 Selection of the Modulation Technique

In Section 5.5.1 it has been explained that a channel does not introduce ISI when the channel coherence bandwidth is larger than the symbol bandwidth. When no ISI occurs, a single-carrier modulation technique can be used without requiring a channel equalizer. This is a cheap and easy solution instead of other more sophisticated approaches, for instance orthogonal frequency division multiplexing (OFDM).

There exists a large variety of digital modulation techniques for single carrier systems, such as phase-shift keying (PSK), amplitude-shift key (ASK), frequency-shift key (FSK), etc. [72]. PSK is known to be the most robust modulation (comparing cases with the same order of modulation) under ideal AWGN channels [72]. That means, that PSK modulation exhibits better BER vs. E_b/N_0[1] curves than the other alternatives.

The wireless channel within a battery is frequency-selective. However, as it was demonstrated in Section 5.5, the coherence bandwidth of the channel is larger than the required symbol bandwidth. Hence, the in-battery channels exhibits similar characteristics to AWGN channels with constant attenuation, due to the almost constant gain of the obtained baseband equivalent channels and the assumption of AWGN. Therefore, PSK modulation is the only one considered in this work.

Binary phase-shift keying (BPSK, $M = 2$) and quadrature phase-shift keying (QPSK, $M = 4$) are the most robust cases among the different modulation orders (M) of the PSK modulation. They have identical theoretical BER vs. E_b/N_0 curves under ideal AWGN channels[72]. Therefore, the transceiver proposed in this work was simulated employing BPSK and QPSK modulations. The results are always practically coincident when the same bit rate is employed, since the coherence bandwidth is in every case large enough. However, QPSK has been the most convenient option for the development of the prototypes presented in Chapter 7. QPSK requires half of the sampling rate for the same bit rate compared to BPSK. This decreases the computational load, which is typically the bottle neck in the employed rapid prototype platforms. Therefore, all the results presented in the current chapter were obtained employing QPSK modulation [104].

The complete simulated communication system consisted of three main blocks: Transmitter, channel (including noise), and receiver. A block diagram of the single-carrier-based communication system is depicted in Fig. 6.2. The channel is composed by the baseband equivalent model of the channel plus the noise generator. The channel is implemented as a FIR filter, whose impulse response is given by its coefficients, the time interval between them, and the number of taps. This impulse response emulates the impulse response of the baseband equivalent model of the

[1] Energy per bit to noise power spectral density ratio.

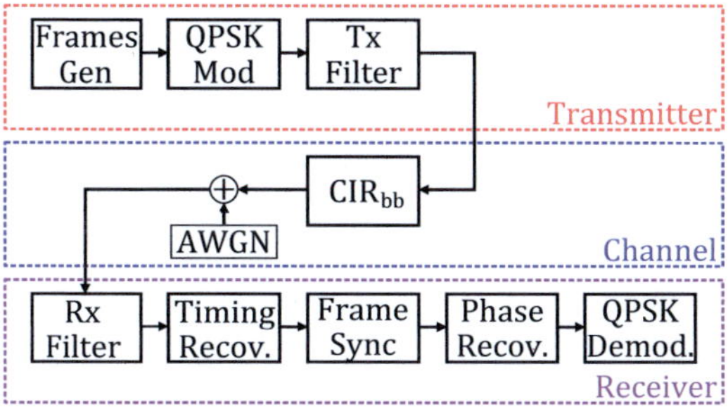

Figure 6.2: Block diagram of the single-carrier-based communication system model implemented for the simulations [104]

channel (CIR_{BB}). The noise is implemented as an AWGN noise block, whose signal is added at the output of the channel. This is the way in which communication systems are mostly modeled in literature [72]. The channel properties and the noise have been already explained in the previous chapter. Therefore, only the transmitter and the receiver are described in detail in the current chapter.

6.1.2 Transmitter

The transmitter is in charge of generating the signals that are injected into the channel. As previously explained, the signal considered at the output of the transmitter in the simulations corresponds to the value that would be given to the DAC in a real digital communication system.

6.1.2.1 Modulation and Filtering of Frames

The communication system under development operates in a frame-based mode. The frame length depends on the communication method (module-wise or cell-wise communication) and in the amount of cells per module in case of module-wise communication. The frame consists of three parts: The preamble, the address and the payload. The preamble is explained in Section 6.1.3.5. The payload contains the information to be sent. The address is used by the MAC layer to recognize the frame recipient. However, the address is not necessary in the simulations performed in the current chapter because the MAC layer is not considered. The frame structure is explained more in detail in Section 6.3.2. For the simulations in this chapter, a frame length of 200 bits (a realistic value for a module-wise communication case) was employed.

The generation of the transmitted signal involves the following steps:

- The information in bits to be transmitted is stored in the payload. In a real application, this would contain the information obtained from the sensors deployed in the cells. In the simulations, the content of the payload is obtained from a pseudo-noise sequence generator.
- The preamble is added at the beginning of the frame.
- The complete frame is modulated (`QPSK Mod` block in Fig. 6.2).
- The modulated frame passes through a transmit filter (`Tx Filter` in Fig. 6.2).

After the last step, the signal is injected into the baseband equivalent model of the channel.

A root-raised-cosine (RRC) filter is employed as transmit filter for pulse-shaping with the aim of minimizing the ISI compared to a rectangular window filter [72]. The transmit filter, which is implemented as a FIR filter, first upsamples the signal and then applies the corresponding filtering. An ideal RRC filter should have infinite length, but the practical implementation has a finite amount of taps. Therefore, it is windowed, which truncates the filter impulse response and introduces a group delay in the transmitted signal of half of the filter length in taps. The transmit filter has three parameters: Roll-off factor (β), length in taps and upsampling factor. The first one is a measure of the excess bandwidth of the filter, i.e. the relationship between the symbol bandwidth (B) and the symbol rate [2] (R_s). This relationship is calculated as [33, 101]

$$B = R_s \cdot (1 + \beta). \tag{6.1}$$

[2] In many books, for example in [26, 72], (6.1) is expressed as $B = \frac{R_s \cdot (1+\beta)}{2}$. But in this work, as already mentioned, the bandwidth is doubled since it is measured from negative frequencies and not from 0 Hz, because the signals under consideration are complex baseband representations.

The second parameter, the length, determines the amount of taps of the FIR filter. The longer this length is, the more similar is the impulse response of the implemented filter to an ideal RRC filter. Finally, the upsampling factor indicates how many samples are delivered at the output of the filter for each symbol at the input, i.e. samples per symbol (sps). The upsampling factor has a direct impact on the sampling frequency of the system (f_s), but it does not modify the symbol bandwidth.

6.1.2.2 Bit Rate

Two different types of bit rates can be defined:

Gross Bit Rate: It is the amount of physically transferred bits per second [30] and does not depend on the frame structure. It can be calculated as

$$R_b = \frac{L_{\text{frame\_b}}}{T_{\text{frame}}}, \tag{6.2}$$

where T_{frame} is the necessary time to transfer a frame, $L_{\text{frame\_b}}$ is the length of the frame, where the sub-index "_b" indicates that the parameter is measured in bits. On the other hand, "_s" means that a parameter is measured in symbols. T_{frame} is the duration of a symbol multiplied by the amount of symbols in a frame, i.e.

$$T_{\text{frame}} = L_{\text{frame\_s}} \cdot T_{\text{sym}} = \frac{L_{\text{frame\_b}}}{\log_2 M} \cdot \frac{sps}{f_s}. \tag{6.3}$$

Then, introducing (6.3) in (6.2) ,

$$R_b = \frac{f_s \cdot \log_2 M}{sps}. \tag{6.4}$$

Net Bit Rate: It is the bit rate of the bits of the frame payload, without taken into account the frame overhead bits [74]. It is obtained as

$$R_{bN} = \frac{L_{\text{payload\_b}}}{T_{\text{frame}}}, \tag{6.5}$$

where $L_{\text{payload\_b}}$ is the length of the payload in bits. Introducing (6.3) and (6.4) into (6.5), the last equation can be rewritten as

$$R_{\text{bN}} = R_{\text{b}} \cdot \frac{L_{\text{payload\_b}}}{L_{\text{frame\_b}}}. \tag{6.6}$$

As it is desired to send the payload information as fast as possible for a given gross bit rate, the relationship $R_{\text{bN}}/R_{\text{b}}$ should be as close as possible to its maximum value, i.e. one.

6.1.2.3 Continuous and Burst Transmission Modes

A frame-based communication system can operate in continuous or burst mode. In the first case, the frames are continuously (successively) sent, without pauses between them. In burst transmission mode, the frames are sent periodically, with a certain pause between them.

Due to the large amount of simulations required in this chapter, the implemented transceiver does not operate in burst mode. The sent information is packed into frames, but these frames are sent continuously without pauses between them, in order to speed up the simulations. Nevertheless, both transmission modes were implemented and tested in the transmitter and receiver models employed for rapid prototype platforms in Chapter 7.

6.1.3 Receiver

The receiver must accomplish the following tasks: signal filtering, synchronization and signal demodulation.

6.1.3.1 Filtering

The first block of the receiver is a matched filter. A filter that is matched to a discrete waveform $s(n)$ has the following impulse response:

$$h[n] = \text{K} \cdot s^{*}[\text{N} - n], \tag{6.7}$$

where K and N are arbitrary constants. Then, the impulse response of the matched filter is a scaled conjugated time-reversed version of the signal from which is matched [72]. The matched filter is the optimal linear filter to maximize the SNR under the presence of white additive noise [87]. The receiver filter should be ideally matched to the received signal, which is affected by the transmit filter and the channel. However, the receiver does not know the CIR_{BB} and it also does not include a channel estimator in order to keep it simpler. Then, the receive filter impulse response is matched to the transmit filter impulse response, because the PSK modulator generates a signal with a constant envelope.

The receive filter is implemented in the form of a FIR filter. As this filter must have causal impulse response, N must be equal or larger than the length of the receive filter impulse response. N is set to be equal to this length to keep it as short as possible. Since the transmit filter is a RRC filter with an odd number of samples and with real values, the taps of the corresponding matched filter are the same as the transmit filter ones. However, the receive filter does not downsample the received signal because the next block (symbol timing recovery) requires an upsampled signal.

6.1.3.2 Symbol Timing Recovery

Since the communication system employs digital coherent PSK modulation, the demodulator block at the receiver must sample the received signal periodically, once per symbol interval, in order to properly recover the information that was transmitted. If the demodulator is not capable of sampling the received symbol at the exact moment, i.e. at the maximum "eye opening", the received information can be strongly corrupted because this is equivalent to ISI [72].

There is always a propagation delay between the modulator at the transmitter and the demodulator at the receiver. A component of this delay is introduced by the channel[3]. But there is another source of symbol timing desynchronization in real digital communication systems: If no external timing reference as a global positioning system (GPS) device is used, the DAC at the transmitter and the ADC at the receiver operates with independent clock signals. These clock signals have in real systems

[3] Also the components of the transceiver introduce a propagation delay. However, this one is normally a multiple of the symbol period, so it must not be corrected by the symbol timing recovery block.

a certain tolerance, i.e. they are not perfectly matched to the desired frequency and they also present a certain drift [17]. The offset between both clock signals also produces an offset in the symbol timing between transmitter and receiver.

The propagation delay is unknown at the receiver and it is also unstable due to the offset between the clock signals. Therefore, the receiver itself must recover the symbol timing from the received symbols. Several digital methods have been developed for coherent single-carrier PSK modulation. In this work, different non-decision-directed timing recovery methods were tested, such as Early-late gate [57], Mueller-Muller [62], Gardner [23], and squaring [64] algorithms. Another alternative is a polyphase filterbank [43], which was not tested in Simulink but in rapid prototyping platforms, as later explained in Section 7.3.4.2. All these methods require each received symbol to be composed by more than one sample. Hence, the downsampling of the signal at the receiver is not realized by the receive filter, but after the symbol timing recovery. All mentioned methods have been tested in the simulation and offered a very similar performance. The results presented in this chapter were obtained employing the squaring method [104].

6.1.3.3 Carrier Frequency and Phase Recovery

In coherent PSK modulation, the information is contained in the phase of each individual symbol. A carrier phase offset between the receiver's carrier and the received symbol's carrier produces crosstalk between the in-phase and quadrature components of the received symbol. Even a very small phase offset could produce a high degradation of the demodulator performance in a noisy channel [72]. Additionally, a carrier frequency offset produces a time-variant carrier phase offset. Therefore, a proper synchronization of the carrier frequency and phase is mandatory when coherent PSK modulation is employed[4].

In real communication systems there are several sources for the carrier desynchronization between transmitter and receiver. One is that the oscillators employed at transmitter and receiver for the carrier generation are

[4] Differential PSK (DPSK) could have been used to avoid the carrier phase recovery block. However, DPSK exhibits a worse performance in comparison with PSK in terms of BER.

independent. Then, they present a limited accuracy and a drift when no external reference is employed, as already explained for the clock signals of the DAC and ADC in Section 6.1.3.2. Another cause is the Doppler effect [82] produced by the relative motion between transmitter, receiver and/or the environment[5]. A carrier phase offset is also introduced by the channel. Even in the case of an ideal channel with constant gain and linear phase, the group delay introduces a constant phase offset between the transmitted and received signals.

No carrier frequency offset was introduced in the simulations presented in this chapter. Therefore, no carrier frequency recovery block was included in the simulation model. However, the phase recovery block is necessary to correct the phase offset introduced by the channel model. The phase recovery method employed in the simulation model is the data-aided maximum-likelihood approach implemented in [57]. This data-aided method requires the use of a preamble, which is explained in Section 6.1.3.5. If a carrier frequency offset would be considered, several techniques are already published for quadrature signals [12, 50, 76]. For example, a Costas Loop is employed in the next chapter in the implementation of prototypes.

6.1.3.4 Frame Recovery

As already mentioned, the communication system for the in-battery communication operates in frame-based mode. Consequently, the receiver must be capable of determining when a frame of information begins along a stream of demodulated data. The end of the frame is then automatically recognizable, since the frame length is known and constant. Again, a data-aided method is employed, taking advantage of the same preamble that is used for the phase recovery. The frame recovery is done by permanently correlating the received symbols with a local copy of the sent preamble.

Since this Simulink model operates with continuous transmission, the design of the frame recovery block is considerably simplified. The distance between subsequent frames is the frame length, and it is known in advance. The frame recovery block stores the received samples in a buffer,

[5] Doppler effect is not considered for the wireless in-battery communication approach, as already clarified in Section 5.2.

whose length is twice the frame length in symbols. When the buffer is full, the correlation between the first half of the buffer content and the local copy of the preamble is performed. The begin of the next received preamble along the buffer is determined by the position of the maximum of the correlation. As the frames are sent subsequently, it is not necessary to compare the maximum of this correlation with a threshold level. After determining the beginning of the frame, the amount of symbols corresponding to the frame length is copied from this starting point and passed to the next block. The first half of the buffer of the frame recovery block is emptied, the second half copied to the first one, and the buffer is again filled with the new received symbols, starting from its second half.

In a burst transmission mode, the result of the maximum of the correlation between the received data stream and the preamble copy must be compared with a certain threshold level. In this way, it can be determined whether a frame of information was really received or only noise from the channel was demodulated while no information was sent. Also an automatic gain control (AGC) would be required as first stage of the receiver chain in this case, in order to guarantee a correlation result independent of the channel attenuation. A more sophisticated frame recovery block, capable of working also in burst mode operation, was implemented for the models employed by the rapid prototyping platforms. This block is later introduced in Section 7.3.4.

6.1.3.5 Preamble

A preamble is a sequence of symbols that is sent at the beginning of each frame and is known in advanced by the receiver. Sequences whose auto-correlation function (ACF) have a similar course as the Dirac delta function, i.e. high peak at $\tau = 0$ and low sidelobes, are good candidates to be used as a preamble [61]. Then, the preamble can be easily detected at the receiver by performing the cross-correlation between the received signal and the reference preamble, as previously explained.

Some classic binary codes that exhibit a proper statistical behavior to be employed as preamble are Barker codes, maximum length sequences (MLS) and Gold Codes. These binary codes originally are sequences containing zeros and ones. When the sequence is composed by -1s and +1s -for example, passing the code through a BPSK modulator- the codes have the property that the peak of their auto-correlation function is equal to their length. Then, the remaining side lobes of the auto-correlation take much smaller values than the peak. The longer the sequence, the larger is the relationship between the highest peak and the largest of the side lobes, making the preamble detection at the receiver more robust. However, following (6.6), the larger the preamble, the lower the relationship between R_{bN} and R_b is, degrading the system performance. Therefore, a compromise must be found for the preamble length.

Barker Codes The Barker codes have the best autocorrelation properties, since they have lower sidelobes that any other sequence of the same lengths. However, sequences of this type only up to a length of 13 are known [4, 7]. This can be short in real communication systems, where a channel with a poor SNR and synchronization impairments can significantly degrade the performance of the correlation for the recognition of the Barker code.

Maximum Length Sequences and Pseudo Noise Codes The MLS are bit sequences generated using linear feedback shift registers (LFSR). Being m the length of the LFSR, the length of the corresponding MLS is $2^m - 1$, because the state with all zeros does not occur [72]. One particular type of MLS are pseudo noise codes (PN-codes). In order to generate a PN-code, the LFSR must have a particular initial state, called generator polynomial, according to its length .

Gold Codes A set of Gold codes is generated by a mathematical combination of two MLS of the same length. A set containing $2^m - 1$ different Gold codes of length $2^m - 1$ is generated with two MLS of length $2^m - 1$. Each of these Gold codes has a good auto-correlation and also a good cross-correlation property against the other codes of the same set. That means, that the maximum of the cross-correlation with any other code is much smaller than the maximum of the auto-correlation. This property of Gold codes can be used for distinguishing the different participants of a communication network by employing different Gold codes of the same length as preambles. This is a big advantage in comparison with PN-codes or Barker codes, since they have only one particular code for each particular length. Gold codes are employed in telecommunications for code division multiple access (CDMA) [25] and also in GPS [69].

Modulation of the Preamble If a BPSK modulation is used, then each bit of the preamble code is converted into one symbol. However, this is not the case for modulations with a higher M. The blocks that employ the preamble, i.e. the phase and frame recovery blocks, operate with symbols and not with bits. But the symbols generated from a modulated preamble with $M > 2$ do not follow the auto-correlation properties expected from the preamble anymore. The adopted solution to this problem is to repeat M-times each bit of the preamble code before its modulation and to employ a Gray code for the symbol mapping in the modulator and demodulator. Thus, every bit of the original preamble is converted into a sub-group of M-times zeros or M-times ones. Each sub-group generates one symbol after the modulator. The possible values which the output of the modulator can adopt for these incoming sub-groups are exactly in opposite positions (i.e. with a phase offset of 180°) of the constellation at the output of the modulator. Then, the obtained preamble symbols after the modulation is always the same as the one of BPSK, independent of the value of M.

A Barker code of 13 symbols was employed for the simulations of the communication system in this chapter, since only one transmitter and one receiver were simulated. Therefore, there is no necessity to distinguish the participants.

6.1.3.6 Demodulation and Frame Decoding

A simple QPSK demodulator with hard decision was employed for demodulation, converting the synchronized symbols into bits. As the frames are already aligned after the frame recovery block, the frame decoder must only discard the first bits of the frame that correspond to the preamble. The remaining bits belong to the payload and are stored.

6.2 Simulation of the Physical Layer Performance

6.2.1 BER vs E_b/N_0

The performance of a communication system is typically measured in terms of the relationship between BER and E_b/N_0. The BER is the ratio between the number of error bits at receiver and the number of received bits. It follows the following equation:

$$\text{BER} = \frac{\text{\#received error bits}}{\text{\#received bits}}, \tag{6.8}$$

where the operator # means "number of".

The transmitted and received bits of the payload must be saved and compared to measure the BER. If the same bit sequence as the Barker code appears by chance in the payload, the frame recovery block could wrongly detect the beginning of the frame. This would significantly increase the measured BER. In order to avoid the explained situation, the payload to be transmitted is not randomly generated. Instead, the payload bits are taken from the first bits of a PN-code which is longer than the payload.

In Simulink the parameter E_b/N_0 can be set directly in the AWGN block, in order to obtain the BER as a function of E_b/N_0. In this Section, the results are presented considering only the case that the optimal carrier frequency has been chosen, following the values of Table 5.2.

6.2.1.1 Results and Discussion

The BER vs E_b/N_0 curves do not reflect the impact of the channel attenuation on the system performance. For example, if the case of two channels with the same CTF form but with different average attenuation[6] is considered, the simulation of the system performance will output the same BER vs. E_b/N_0 curve with both channels.

As explained before, the bandpass channels obtained with the optimal center frequencies are almost flat in the considered bandwidth. Consequently, all the obtained BER vs. E_b/N_0 curves are very similar to the

[6] In a |CTF| in dB vs. frequency plot, this situation would be observed as two similar traces but with a vertical offset.

theoretical curve of QPSK within an AWGN channel, since the communication system does not suffer from ISI. As an example, Fig. 6.3 shows the E_b/N_0 curves of the worst cases with PIFA and helix antennas under the architecture 1M-4S.

The BER vs. E_b/N_0 curves are normally useful, for example, for the evaluation of the influence of one receiver component (e.g. different synchronization techniques) or an impairment in the communication system (e.g. a frequency offset between transmitter and receiver). However, these curves alone do not provide much useful information if we compare the performance of the system in different channels, because different transmit signal power is employed in each case, in order to obtain the same E_b/N_0 value at each receiver. Therefore, the next section introduces a new test procedure which was employed to analyze the performance of the proposed system.

6.2.2 BER vs. Transmission Power

It was proved with the channel characterization and the BER vs. E_b/N_0 curves that a single-carrier system with QPSK modulation and the proper synchronization techniques is capable of a good quality transmission at values of E_b/N_0 below 10 dB, even without employing a channel equalizer at the receiver. This accomplished the first goal of this chapter, that was the design of the transmitter and receiver architectures. The second aim

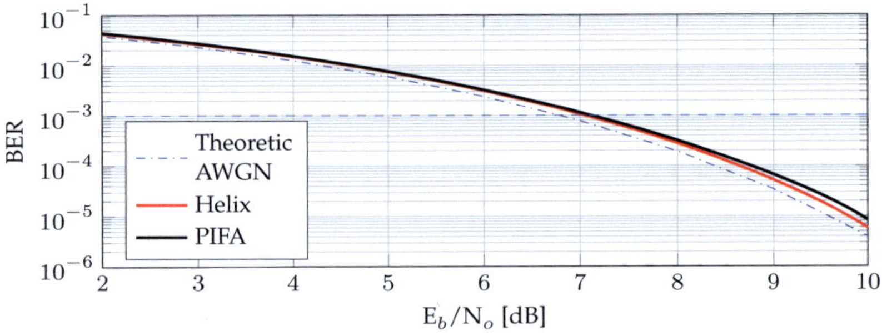

Figure 6.3: BER vs. E_b/N_0 curves for the worst case of the simulated channels with helix and PIFA in the battery emulator, with the architecture 1M-4S

was to estimate the necessary signal power at the transmitter, in order to ensure the minimum required BER.

The transmission power (P_{Tx}) for a particular energy pro bit at the transmitter is obtained as

$$P_{\text{Tx}} = E_{\text{b(Tx)}} \cdot B_{\text{sym}} = E_{\text{b(Tx)}} \cdot \frac{f_{\text{s}}}{sps} \tag{6.9}$$

where $E_{\text{b(Tx)}}$ is the energy per bit at the transmitter and, as previously said, f_{s} is the sampling frequency and sps the amount of samples per symbol. Then, by knowing the CTF and the architecture of transmitter, the relationship between $E_{\text{b(Tx)}}$ and $E_{\text{b(Rx)}}$ is [104]

$$E_{\text{b(Tx)}} = \frac{E_{\text{b(Rx)}}}{\frac{1}{K^2} \cdot \sum_{k=0}^{k=K-1} \left| H_{\text{TxF}}[k] \cdot H_{\text{ch}(i,l)}[k] \right|^2} \tag{6.10}$$

where $E_{\text{b(Rx)}}$ is the energy per bit at the receiver (also known simply as E_{b}), $H_{\text{ch}(i,l)}[k]$ is the discrete channel frequency response, $H_{\text{TxF}}[k]$ is the discrete frequency response of the transmit filter, and K is the length of the $H_{\text{TxF}}[k]$ and $H_{\text{ch}(i,l)}[k]$[7]. Taking (6.10) into account, the BER vs. E_{b}/N_0 can be easily converted into BER vs. $(E_{\text{b}}/N_0)_{(\text{Tx})}$ [8] curves by dividing the values of the E_{b}/N_0–axis by the denominator of (6.10). When many curves obtained from different channels are plotted together into the same graph the conversion of horizontal axis values must be carried out independently for each curve, according to the corresponding channel. This displacement depends on the channel attenuation: A vertical displacement in the CTF in decibels produces the same displacement in decibels in the horizontal direction of the BER vs. $(E_{\text{b}}/N_0)_{(\text{Tx})}$ graph.

6.2.2.1 Results and Discussion

Following the procedure explained below, the BER vs. $(E_{\text{b}}/N_0)_{(\text{Tx})}$ curves were obtained for each of the four simulated architectures with both types of antennas. Due to the large number of results, only the best and worst cases for each architecture and antenna type are plotted.

7 If $H_{\text{TxF}}[k]$ and $H_{\text{ch}(i,l)}[k]$ have different lengths, the shorter must be interpolated in order to have the same length of the largest.

8 Energy per bit at the transmitter to noise power spectral density ratio.

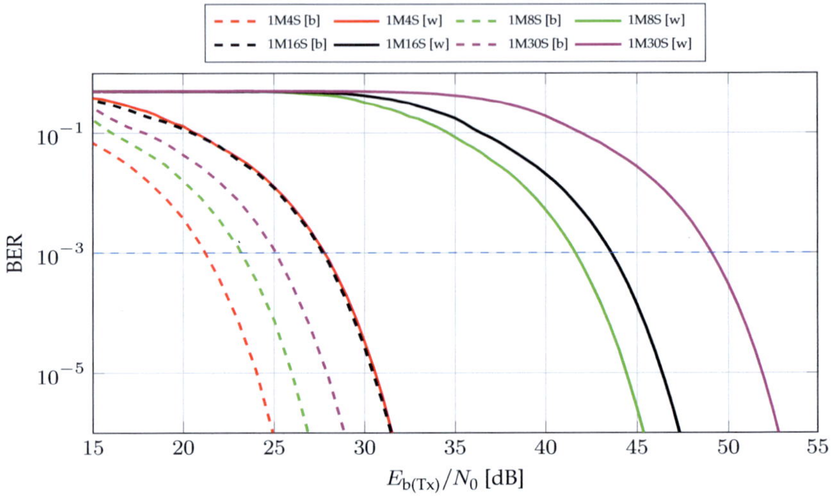

(a) Helix antennas, first frequency sub-range

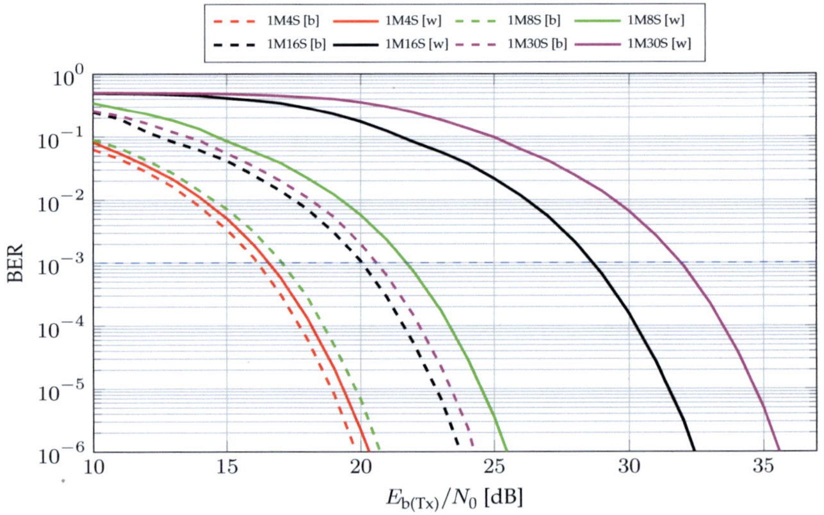

(b) PIFAs, third frequency sub-range

Figure 6.4: BER vs. $(E_b/N_0)_{(Tx)}$ curves simulated for the channels with helix or PIF-Antennas in the small battery emulator. References: (b) = best case, (w) = worst case

The results are shown in Fig. 6.4(a) for the helix antenna in the first frequency sub-range, and in Fig. 6.4(b) for the PIFA in the third frequency sub-range. The difference in performance when employing different antennas (at different frequencies) and when changing the amount and/or position of this antennas is clearly observable with the BER vs. $(E_b/N_0)_{(Tx)}$ curves. Some conclusions can be drawn by analyzing the results:

- In most cases, a higher $(E_b/N_0)_{(Tx)}$ is required to achieved the same BER when the amount of antennas increases, i.e. the system performance decreases. The reason is that the energy of the transmitted signal is radiated along the entire emulator and shared by all antennas.
- The performance at higher frequencies is better than at lower frequencies, when the cases with the same amount of antennas are compared. The helix antennas required in every architecture an $(E_b/N_0)_{(Tx)}$ between 5 dB and 20 dB higher than the PIFAs to achieve the desired BER$< 10^{-3}$. The $(E_b/N_0)_{(Tx)}$ required in the worst case of all scenarios is 31.95 dB with PIFAs and 49.10 dB with helix antennas.
- The difference in $(E_b/N_0)_{(Tx)}$ between the best and worst case is bigger with helix antennas in most scenarios.

The required $(E_b/N_0)_{(Tx)}$ in every case to achieve the BER of 10^{-3} can be obtained from the curves in Fig. 6.4(a) and 6.4(b). Then, following (6.9), the power transmission can be also obtained as

$$P_{Tx} = (E_b/N_0)_{(Tx)} \cdot N_0 \cdot \frac{f_s}{sps}. \tag{6.11}$$

N_0 was calculated with (5.1) in Section 5.3. Table 6.1 shows the results.

Since the same N_0 value is used in both frequency sub-ranges, the same conclusions as from the BER vs. $(E_b/N_0)_{(Tx)}$ curves can be drawn for the required transmission power, when comparing the values of required transmission power for different amounts of antennas and different frequency sub-ranges. But additionally, these results prove that the required transmission power is very small in every case: The worst case at lower frequencies is 363 pW (–64.41 dBm) and 6.98 pW (–81.56 dBm) at higher frequencies.

Table 6.1: Necessary $(E_b/N_0)_{(Tx)}$ and P_{Tx} for a BER $= 10^{-3}$ (without channel coding).

Freq. sub-range	Archi-tecture	$(E_b/N_0)_{(Tx)}$ [dB]		P_{Tx} [dBm]	
		best	worst	best	worst
Helix	1M-4S	21.36	27.82	-92.15	-85.68
	1M-8S	23.29	41.66	-90.21	-71.84
	1M-16S	27.82	43.63	-85.68	-69.87
	1M-30S	25.16	49.09	-88.34	-64.41
Third (PIFA)	1M-4S	16.22	16.69	-97.28	-96.81
	1M-8S	17.10	21.82	-96.41	-91.68
	1M-16S	20.09	28.77	-93.41	-84.73
	1M-30S	20.73	31.94	-92.78	-81.56

6.3 Description of the MAC Layer

6.3.1 Channel Access Method: TDMA

The channel access method is the technique employed to allow many devices to transfer and/or receive information over the same shared communication medium. There are several types of channel access methods, but most of them can be placed in one of the four next categories [22, 58]:

- Time division multiple access (TDMA): It is based on time-division multiplexing (TDM). The communication process is organized in cycles with a particular duration. Each transmitter has a particular time slot assigned along this cycle, in which it must transmit its information. Consequently, the transmitters are not allowed to transfer information at the same time.

- Frequency division multiple access (FDMA): It is based on frequency-division multiplexing (FDM). The complete spectrum allocated for the communication is divided into narrower frequency sub-bands, which are assigned to the different transmitters. The transmitters can send information at the same time.

- Code division multiple access (CDMA): It is based on spread spectrum. The transmitters share the frequency band and are allowed to transmit simultaneously by employing different orthogonal spreading codes.
- Space division multiple access (SDMA): Each transmitter transfers information in different physical areas or in different physical directions.

When deciding the channel access method for the IntLiIon project, SDMA was automatically discarded, since the master and slaves share the same physical area: The battery. TDMA is the access method employed in the state-of-the-art BMSs that employ CAN bus. In order to allow an easier transition between the current implemented technologies and the new ones coming from the IntLiIon project, it was decided to continue employing TDMA as channel medium access technique. A scheme of TDMA is shown in Fig. 6.5.

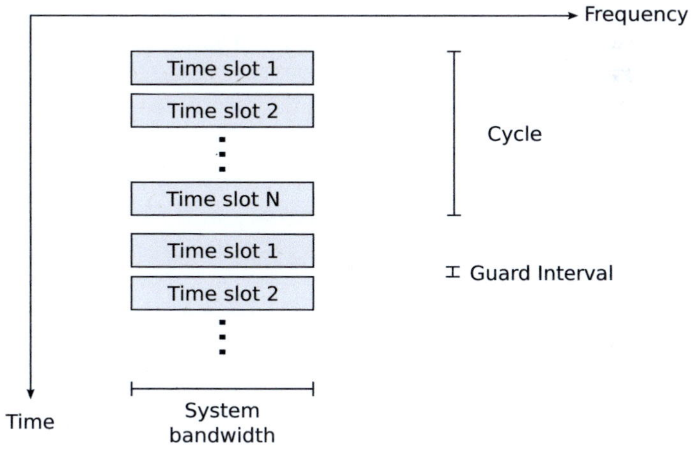

Figure 6.5: Scheme of TDMA [36]

6.3.2 Communication Protocol

The battery communication network has a master-slave architecture, as already explained in Section 1.2, where the BCU is the master and either the modules or the cells are the slaves. Each communication cycle must begin with a requirement from the master, to which each slave must respond in its assigned time slot. The information that the slaves send to the master in the battery communication network is always the same: Temperature and voltage data from the cell sensors[9]. The periodicity with which this information must be transferred is also constant and it must be kept always shorter than the latency requirements given in Section 1.4.

The two physical variables that must be transmitted, temperature and voltage, have two clearly different latency requirements: 100 ms for the first one and 10 ms for the second one. Transferring both variables every 100 ms would not meet the requirement on the voltage latency. Sending this information every 10 ms would mean a very large amount of unnecessary information, because the variation of the temperature with the time is much slower. Hence, the communication process in the battery network in this wireless approach is organized in cycles and sub-cycles [106]. Along one sub-cycle, the BCU (master) must receive the voltage information of every cell in the battery. Therefore, the maximum duration of one sub-cycle is 10 ms. Along one cycle, the BCU must receive the temperature information of every cell, so the duration of a cycle is limited to 100 ms.

The relationship between the duration of a cycle and a sub-cycle is a factor of 10, similar to the number of cells that one module typically contains. Hence, the adopted solution was to send in each sub-cycle the voltage information of every cell in the battery, and the temperature information of one cell per module. The cell in the module whose temperature is transferred varies in each sub-cycle.

N_{cm} being the number of cells in a module and N_{m} being the number of modules in the battery, in each sub-cycle the voltage values of $N_{\mathrm{m}} \cdot N_{\mathrm{cm}}$ cells (the total amount of cells in the battery) and the temperature of N_{m} cells must be sent. Then, a cycle must be composed of N_{cm} sub-cycles. After a complete cycle, the BCU receives the temperature values of the

[9] More information is expected to be transferred in a real automotive battery. For example, emergency messages. However, this was not considered in the IntLiIon project.

$N_m \cdot N_{cm}$ cells in the battery. Between time slots there is always a time guard interval to avoid the overlapping between frames transmitted in subsequent time slots, in case of timing offsets between the communication devices.

The information that each communication device (master or slave) transmits along a time slot receives the name of frame. However, the information transferred by master and slaves is not the same.

Master Frame It was decided that the first time slot of the every communication sub-cycle is assigned to the master, which sends a frame addressed to all slaves, i.e. a broadcast frame. This frame is intended to be used by the slaves for synchronization: They know in advance the time slot that has been assigned to them, so this time slot is used as an offset value from the beginning of each sub-cycle. This would solve problems with small timing offsets between the different slaves. This is important, since the BMS must be able to operate continuously for long periods of time. The master frame must not actually transfer any information to the slave. It must be only detected. Therefore, this frame contains only a preamble.

Slave Frame The slave frames have the following structure:

- Preamble: It is employed for the frame synchronization and as training sequence for the estimation of timing and phase offsets, as explained in Section 6.1.3.5
- Address: It is used in the MAC layer for identifying the slaves
- Payload: It represents the information transmitted, i.e. voltage and/or temperature of cells

Regarding the identification of the slaves, other two alternatives to distinguish the transmitter of each frame instead of the address could be that every slave uses a different preamble or checking the time slot at which each frame was received at the master.

Cell-Wise Communication In a cell-wise communication, there must be a total of $N_\mathrm{m} \cdot N_\mathrm{cm} + 1$ time slots. $N_\mathrm{m} \cdot N_\mathrm{cm} - N_\mathrm{m}$ slaves transfer only the voltage values in their time slots. The length of each of these frame's payloads is 14 bits. The remaining slaves (N_m) send both voltage and temperature values, employing 25 bits for the payload.

Module-Wise Communication There are $N_\mathrm{m} + 1$ time slots per sub-cycle in a module-wise communication. Each of the N_m slaves transfers the N_cm voltage values and one temperature value in each frame. Then, the length of the frames is $(14 \cdot N_\mathrm{cm} + 11)$ bits.

In the IntLiIon project it was decided to give priority to the module-wise communication, since less modems are required. These devices would be even simpler, because the communication channel is less frequency-selective with a smaller number of antennas. Therefore, only a module-wise communication is implemented by means of rapid prototyping platforms in Chapter 7. The communication protocol for module-wise communication is shown in Fig. 6.6.

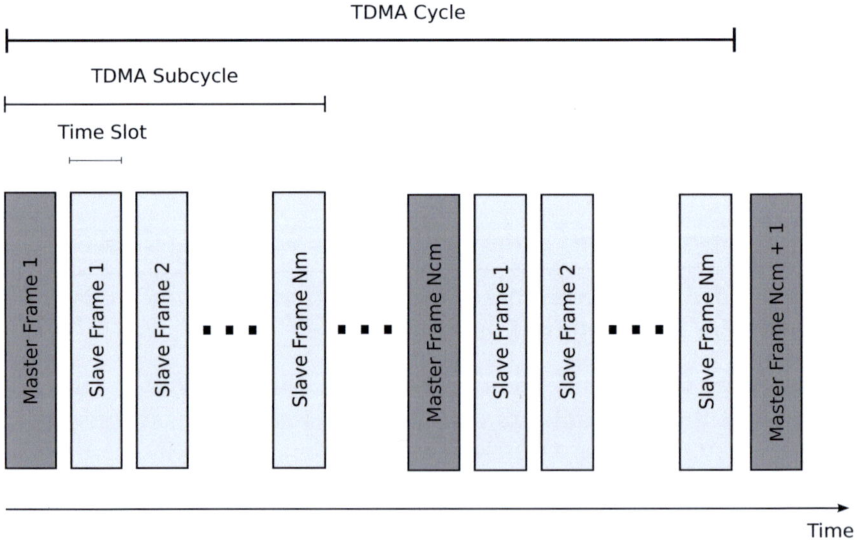

Figure 6.6: Scheme of the communication protocol for module-wise communication

7 Prototyping of the Communication System

One of the main goals of the IntLiIon project was the development of a demonstrator for the wireless data transmission within an automotive battery. The demonstrator must be composed by the propagation environment represented by the same testbed (channel emulator plus antennas) as employed for the measurements, as well as the prototypes of the modems used for the wireless communication. The testbed was already explained and employed in previous chapters and, therefore, the current chapter focuses on the implementation of the modem prototypes. The chosen rapid prototyping devices for the development of the modems were off-the-shelf software-defined radio (SDR) platforms.

This chapter starts with a brief introduction to the topic of prototyping by means of SDR platforms and explains which platform models and software framework were employed. Then, it concentrates on the explanation of the complete implementation of the physical and MAC layers of the communication system, as well as the performance evaluation carried out by means of simulations and measurements.

7.1 Prototyping with Software Defined Radio

7.1.1 Software Defined Radio Platforms

An SDR platform is a flexible radio frequency hardware that performs the typical tasks required to convert a high frequency analog signal into a digital baseband signal and/or vice versa. Basically, an SDR platform carries out in hardware the tasks of the digital transceiver described in Section 6.1, with exception of the baseband signal processing. This is carried out by an embedded system, such as a microprocessor, a field programmable gate array (FPGA), or a digital signal processor (DSP). This embedded system is either included in the SDR platform or in a host computer connected to the SDR platform. An ideal SDR platform should be able to operate over a large range of frequencies, possess a flexible bandwidth, transmit and receive signals within a wide dynamic range, while maintaining spurious-free operation [59, 63].

7.1.2 Universal Software Radio Peripheral (USRP)

The Universal Software Radio Peripheral (USRP) is a family of software-defined radio platforms [89]. It can be programmed with different environments such as GNU Radio, LabView or Matlab/Simulink. GNU Radio was employed in this work, since it was not possible to operate under real-time conditions at the desired sampling rate with commercial computers under Labview or Matlab/Simulink.

The model employed during this work for the prototyping of the communication system was the USRP N210. This model operates together with a host computer. The main electronic board of the USRP, called motherboard, includes an FPGA, which controls the ADC and DAC, and handles the communication with the host computer via Ethernet. The host computer and the FPGA interchanges digital samples in baseband as complex values. The USRP N210 must be equipped with the so-called daughterboards, which are the front ends needed for down- and up-conversion and filtering. By using different daughterboards, the USRP can operate in different frequency ranges from 0 Hz to 6 GHz. Some of the most important technical characteristics of the N210 are [75, 89]:

- FPGA: Xilinx Spartan 3A-DSP 3400
- 100 MSps dual ADC
- 400 MSps dual DAC
- Stream transfer up to 50 Msps to and from host computer with 8-bit samples, or up to 25 Msps with 16-bit samples.

The driver employed as interface between the USRP and the host computer is called USRP Hardware Driver (UHD).

7.1.2.1 Burst Transmission with USRPs

Not only the physical layer but also the MAC layer were developed with USRPs. These devices offer a simple alternative to operate in burst transmission mode, which is required for the implementation of the TDMA protocol for the MAC layer. This burst transmission is controlled by means of special stream tags.

Tag: A stream tag, also known only as tag, is a piece of metadata that is transmitted together and synchronized with a sample. Tags propagate through the flow graph between blocks and always in the same direction as the streams [52, 53, 88]. A tag can contain several fields with different information. Some examples of these fields are [53]:

- Key: Used by the blocks for the identification of the tag
- Source: Specifies the block that originated the tag
- Offset: Specifies the sample position of the tag in a stream
- Value: Contains a certain piece of information that is transferred

Local Time of the USRP: A USRP must accurately know its local time for a burst transmission in order to send the burst at the desired moment (time slot). This can be done based on a special types of tags included in the UHD and generated by the USRP Source and USRP Sink blocks. When a USRP Source starts a stream transfer, it sends an rx_time tag together with the first stream. This occurs when a flow graph starts its execution or after a change in the center frequency of the USRP. This tag contains the current local time of the USRP associated to this USRP Source when this first stream is received. Then, as the sampling frequency of the USRP is known, the local time of the USRP can be updated when a new stream is received by the USRP Source. The product of the sampling time and the amount of received samples must be added to the last calculated time. Another possibility to obtain the local time of the USRP is to synchronize it with its host computer's local time. A problem with this alternative arises when many USRPs must operate synchronized but each of them is controlled by a different computer. The computers can be synchronized by the network time protocol (NTP). However, the inaccuracy of this protocol, typically in the order of the milliseconds, is extremely big compared to the symbol duration of the transmitted signals. Therefore, the first alternative, based on the calculation of the USRP's local time by means of the rx_time tag, is preferred.

Burst Transmission: A data stream can be transmitted as a burst by an USRP Sink at a particular instant of time when the stream contains the following tags:

- tx_sob[1]: It is placed at the first sample of the burst
- tx_eob[2]: It is located at the last sample of the burst
- tx_time: It is also placed at the first sample and indicates at which time the burst must be sent

When the USRP Sink receives a stream with these tags, the associated USRP starts to work as transmitter at the time indicated by tx_time. When the sample with the tag tx_eob is received by the USRP Sink, the transmission stops and the USRP changes to receiver mode. That means, that the USRP operates always as receiver, except when it must send a tagged burst. In this way, the USRP can work very easily in a half-duplex way. When the WBX (the one used in the demonstrator) or SBX daughterboards are employed, the port used for this operating mode is the Tx/Rx port, because the Rx2 port can operate only as receiver [89].

7.1.3 GNU Radio

GNU Radio is a free software development toolkit licensed under the GNU General Public License (GPL). It provides signal processing runtime and processing blocks designed for the implementation of signal processing systems and software defined radios [78]. It can be used only as a simulation environment or, combined with the appropriate hardware platforms, for fast prototyping of software defined radios. The blocks can be connected between them, forming signal flowgraphs that can be executed as programs. The flowgraphs, as well as the signal processing blocks, can be written in Python or C++. The last one is normally preferred for time-critical signal processing tasks. A graphical user interface, called GNU Radio Companion, is offered for simplifying and boosting the design of signal flowgraphs with GNU Radio and generating flowgraph source code [28].

---

1 sob: start of burst

2 eob: end of burst

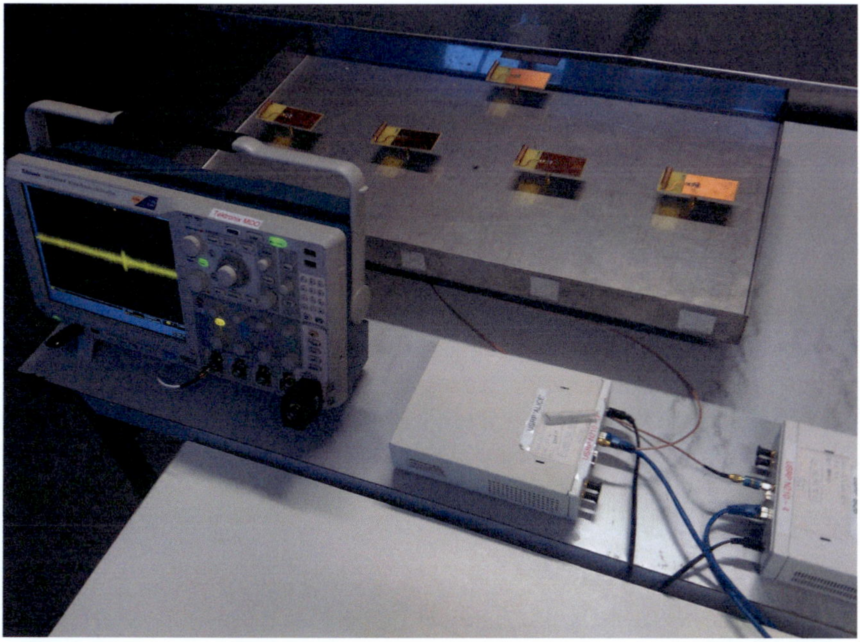

Figure 7.1: Photo of the demonstrator, with the emulator with the top cover removed

7.2 Demonstrator of the Wireless In-Battery Communication

The aim of using rapid prototyping platforms was to test the proposed communication system in a real environment, instead of only by means of simulations. As in the previous chapters, the small channel emulator was employed as testbed for the demonstrator. The antennas employed were the customized helix versions under the configuration 1M-4S. The same enumeration of the slaves applies for this demonstrator. By using these antennas, it was possible to measure the signals inside the emulator in the time-domain employing an off-the-shelf mixed domain oscilloscope (MDO), which has a maximum bandwidth of 1 GHz. Fig. 7.1 shows a photo of the demonstrator without the top cover. As before, the emulator was always closed during the measurements.

The CTFs between master and slaves were measured with a VNA. They are shown in Fig. 7.2. The optimal center frequency obtained by means of the algorithm from Section 5.4 was 402.5 MHz.

During the measurements, two of the antennas were connected to USRPs emulating the communication modems. One USRP was the master and the second one acted as slave, connected to the position S2. The slave antenna S1 was connected to a MDO, in order to observe the signals inside the emulator either in time or frequency domain. In some measurements, the slave S4 was connected to a third USRP in order to test the system with a second slave at the same time. The antenna in the position S3 was terminated with a 50 Ω load, as well as the antenna in the position S4 when no third USRP was employed.

The results presented in this work with the demonstrator were obtained with antennas for the first frequency sub-range, i.e. for frequencies below 700 MHz. Therefore, the USRPs were equipped with the WBX daughterboards. These allow an operating frequency range from 50 MHz to 2200 MHz with a maximum bandwidth of 40 MHz [89]. The demonstrator could also operate in the third frequency sub-range employing the SBX daughterboards (from 400 MHz until 4400 MHz).

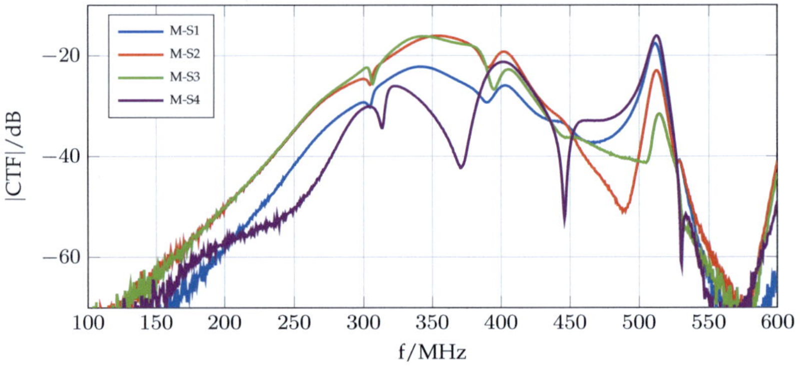

Figure 7.2: Measurement of the CTFs between master and slaves in the small emulator with the customized helix antennas

7.3 Implementation of the Physical Layer

For the implementation of the physical layer without the MAC layer, the communication system consists of three main components: The transmitter, the channel (which includes the noise source) and the receiver. Fig. 7.3 shows a block diagram of the implementation. There is one shaded component: The Burster. This is only required in burst transmission mode, but not in continuous transmission mode.

The transmitter and receiver are explained in detail below. First, only the case without channel coding is explained. Then, channel coding and its two components, encoder and decoder, are separately explained in Section 7.5.

7.3.1 Modem configuration

As in the system design for simulations in Section 6.1, QPSK was employed as modulation technique. If the amount of samples per symbol would be set to four, the sampling frequency of the USRP should be set to 4 MHz for a bit rate of 2 Mbps. However, 4 MHz is not a valid sampling frequency for the USRP. Therefore, a sampling frequency of 5 MHz was chosen. In order to obtain the desired bit rate of 2 Mbps, a upsampling factor of five was set at the transmitter, and the same value for the downsampling factor at the receiver. The center frequency of the both transmitter and receiver was set to the optimal value, i.e. 402.5 MHz.

7.3.2 Transmitter

The digital transmitter is responsible for generating the information frames, as well as the modulation and pulse shaping of the signal to be transmitted.

7.3.2.1 Frames Generator

The frame structure employed for the prototyping of the physical layer is the same as the one explained for frame slaves in module-wise communication in Section 6.3.2, since only module-wise communication is considered for prototyping.

Preamble: A length of 64 symbols (128 bits) was empirically chosen for the preamble. Through simulations it turned out that a preamble of 32 symbols or less was not enough to ensure a proper performance, since the amount of lost frames increased considerably. A preamble longer than 64 symbols did not improve the performance.

Gold codes were chosen as preambles, allowing the use of different codes but with the same length. The implementation of TDMA takes advantage of this issue in order to distinguish the master frames from the slave frames. This decreases the possibility that a slave confuses the beginning of a slave frame with the beginning of a master frame, which represents the start of a new sub-cycle.

Address: An address of one byte allows the generation of 256 different addresses. Since the amount of modules in a battery is typically much smaller, a reduced set of valid address can be employed but with the possibility to increase the Hamming distance between them.

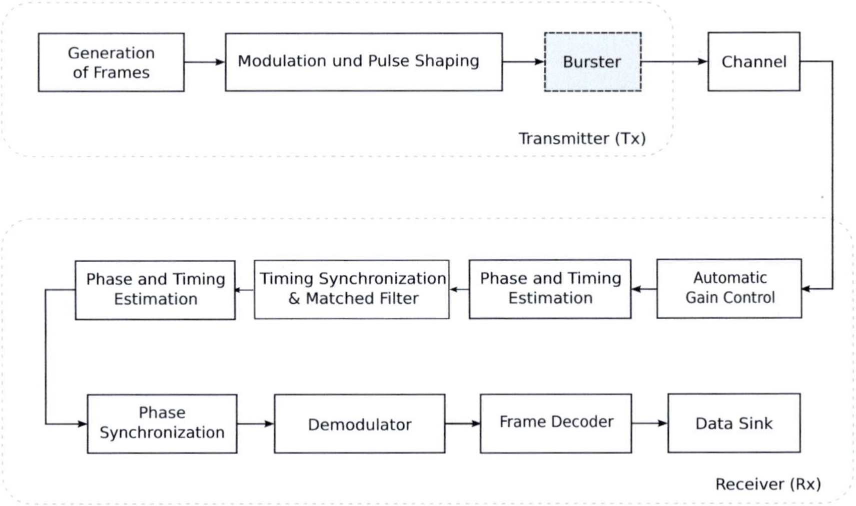

Figure 7.3: Scheme of the physical layer implementation for continuous (without Burster) and burst transmission modes, without channel coding

Payload: The length of the frame's payload is $(14 \cdot N_{cm} + 11)$ bits. N_{cm} is typically between 8 and 14. That means, that the length of the payload would be between 123 and 207 bits, i.e. 16 and 26 bytes. A length of 20 bytes was employed. In this implementation, to the case of the Simulink model, the payload information was generated by means of a PN-code.

The complete frame is generated with a Vector Source block in GNU Radio. A Head block is also used, in order to terminate the execution of the flow graph after a fixed amount of frames. This is important for the automation of the BER measurements. Otherwise, the flow graph executes continuously until it is manually terminated.

7.3.2.2 Modulator and Pulse Shaping

In GNU Radio, both modulation and pulse shaping are carried out by the block Constellation Modulator. There, QPSK was set as modulation technique, Gray code as mapping method, and RRC as filtering technique, with an upsampling factor of five and a roll-off of 0.35.

7.3.2.3 Burster

The Burster is only necessary for the burst transmission mode. This block must be removed in case of continuous transmission. Two different alternatives were designed as Bursters. The first and simplest one has been only employed in simulations of the physical layer. The second one works only together with a transmitter USRP. In every case, each sent burst stores the information corresponding to one frame.

Burster for Simulations: This Burster consists of a multiplexer (Mux block in GNU Radio). This block has two inputs for streams, which are serially concatenated at its output. The amount of samples to be taken from each input for the generation of an output stream is set individually. The data frame is connected to one input. A Null Source, which continuously outputs a stream containing zeros, is connected to the other input. The amount of zeros that is placed between successive frames is chosen according to the desired time separation between frames.

Burster for Measurements with USRPs: The task of the Burster is to properly tag each stream with the tags explained in Section 7.1.2: tx_time, tx_sob, and tx_eob. Two different blocks were designed to act as Bursters: Master Frame Tagger and Slave Frame Tagger. The first one is employed as Burster in the transmitter in burst mode of the physical layer, as well as in the master in the TDMA implementation. The second one is the Burster in the slaves in TDMA. Therefore, only the Master Frame Tagger is explained in this section. The Slave Frame Tagger is explained in Section 7.6.3.

Master Frame Tagger: The parameters to be set in this block are the length of the stream to be transmitted and the transmission periodicity. Every time that a frame is transferred, the Master Frame Tagger block tags the next frame. The tx_sob and the tx_eob tags are placed at the start and end of the new frame, respectively. The addition of the time at which the last frame was sent and the time interval between successive frames is loaded into the tx_time tag.

The Master Frame Tagger is based on the Packed Pad and Burst Tagger blocks from the out-of-tree module *gr-foo* [5]. The Burst Tagger block looks continuously for a specific tag, whose name (Key) must be set in the block properties. When this tag is found, the Value of this tag corresponds to the burst length, and its position (obtained by the offset field) is the beginning of the burst. Then, the Burst Tagger tags the burst with the tx_sob and the tx_eob at the corresponding samples. Packed Pad is a block capable of adding a tag at a particular sample along a stream. For the implementation of the Master Frame Tagger, the Burst Tagger block was used as starting point, but its source code was modified by adding a function (copied from the source code of the Packed Pad to add a tx_time tag to the first sample of the burst, together with the tx_sob tag.

The Master Frame Tagger was implemented as a synchronous block. That means, that the sampling rate of the output signal is the same as for the input signal. The following parameters must be set in the Master Frame Tagger block:

- Tag Name: ID of the tag that is searched for at the beginning of the frame and whose value parameter contains the length of the burst

- Multiplier: This value indicates the amount of samples per bit. The length of the burst would be the product of the value of the Multiplier and the value parameter of the tag at the beginning of the frame. In the designed model, this value is one
- Time between Bursts: The time (period) between two subsequent bursts, in seconds
- Initial Delay: The time which passes until the Master Frame Tagger sends the first burst.

The Master Frame Tagger, as well as the Slave Frame Tagger, cannot be used in a simulation in GNU Radio without the USRPs, since its functioning requires tx_time tag, which is generated by the UHD.

7.3.3 Communication Channel

Two different cases must be considered for the channel: When the system is only simulated (without USRPs) and when the system is tested with USRPs. In case of a simulation, a baseband equivalent model of the channel is employed for simulating the effects of the channel. The channel is implemented by a FIR filter in a similar way as in Simulink, employing the Channel Model block from GNU Radio. The coefficients of this channel model are also obtained in the same way as for the Simulink model. However, in this case the baseband equivalent was obtained from the channel measured in the emulator with the VNA.

The small channel emulator with five self-costumized helix antennas was employed for the performance measurements with the USRP, as previously explained in Section 7.2.

A frequency offset of around 30 Hz was measured between the SDR platforms oscillators. This value was employed for the simulation of the system in GNU Radio, also by using the Channel Model block.

7.3.4 Receiver

The digital receiver reads the sampled signal, converts it into bits, and decodes the received frames.

7.3.4.1 Automatic Gain Control

The AGC is necessary because the next block requires a constant signal power at its input for an appropriate functioning. At the moment of the development of this model, there were several AGC models available in the libraries of GNU Radio. Most of them presented a good performance in a continuous transmission mode, but not under burst transmission. A new AGC block was developed, employing a moving average algorithm. This block performed properly in both transmission modes.

7.3.4.2 Timing and Carrier Synchronization

As it was explained in Sections 6.1.3.2 and 6.1.3.3, there are several reasons for a timing and phase desynchronization. The proper timing and phase of the received symbols must be recovered in order to ensure a good performance.

There are several ways to correct these offsets very accurately with the USRPs. For example, by means of GPS modules or a physical wired connection between the platforms with an off-the-shelf so-called multiple input multiple output (MIMO) cable [20]. Nevertheless, these alternatives were not considered for the prototyping of the communication system. GPS modules could not be employed inside an automotive battery, due to the shielding of the metallic housing, and also because the costs would increase. A physical connection between the communication modems for the synchronization would be also completely senseless, because avoiding additional wiring is the main target of the IntLiIon project. Therefore, the synchronization of the platforms must be done in the digital domain, i.e. at software level.

Blocks for Timing and Carrier Synchronization: The timing and carrier synchronization is implemented by two blocks already available in GNU Radio: Polyphase Clock Sync [31, 43] for the timing and Costas Loop [12] for the carrier. Both blocks are capable of estimating and correcting the timing and phase offset of the received signal.

However, both blocks require a certain amount of samples to converge to an offset estimation. Since the lack of synchronization between SDR platforms is bigger than between transmitter and receiver in the simulations due to the already explained reasons, the preamble must be longer than in the simulations of Chapter 6. This is particularly important in burst transmission mode: If the preamble is not long enough, then the system would not be yet properly synchronized when the first samples of the payload are received. This would lead to a decrease in the system performance.

In order to accomplish a faster estimation of the timing and phase offsets and decrease the required length of the preamble, a Correlation Estimator block can be employed. This block is placed before the synchronization recovery blocks and it correlates the received streams with a copy of the transmitted preamble. The absolute value of the correlation is normalized. One is the maximum value and null the minimum value. A threshold value must be set between these two values. When the absolute value of the correlation's maximum peak exceeds the threshold, the block assumes that a preamble was detected. The Correlation Estimator calculates the timing and phase offset between the received and the transmitted preamble. This information is added by means of tags to the first sample of the detected frame. Next, the Polyphase Clock Sync and the Costas Loop use the estimated timing and phase offsets, respectively, as initial conditions. In this way, both can converge faster to an accurate estimation.

Issue with the Correlation Estimator: It has been observed in [106, 119] that without the Correlation Estimator and with the frame length of 128 bits, the system can still work in continuous transmission mode. However, the first frames are almost completely wrongly decoded, because the offset-estimation does not converge in time. On the other hand, the frames are always completely incorrectly decoded without the Correlation Estimator in burst transmission mode: After one frame is received, there is only noise in the channel and the synchronization blocks are completely incapable of reaching a good estimation along the preamble length. Thus, the first samples of the payload are not properly synchronized. As the final target of the communication system prototype is to implement a TDMA system based on burst transmissions, the Correlation Estimator must be always included.

An important issue with the Correlation Estimator was observed: When the tag produced by the Correlation Estimator propagates through the

Polyphase Clock Sync, which downsamples the signal, the tag does not propagate anymore in the correct location, i.e. at the beginning of the frame. The offset between the new position of the tag and the right one is normally only a small amount of samples but it varies from frame to frame. Hence, it cannot be compensated by a constant shift of the frame. Due to the offset, the Costas Loop applies the phase correction to the frame from a wrong sample position, which is equivalent to a time shifting in the arrival of the frame. This produces an offset in the estimated phase. The Costas Loop is capable of correcting this phase offset but it cannot solve the $\frac{\pi}{2}$ - ambiguity of the QPKS modulation anymore. The solution proposed in [106] is to employ a differential enconding in the modulation, i.e. differential QPSK (DQPSK) instead of QPSK. The differential enconding produces, however, a decrease in the BER performance of the system [72]. In the case of the continuous transmission, there is an additional consequence of this wrong tag position: The correction of the Costas Loop block is not applied within one frame, but also on some symbols in the previous or the next frame. All these last symbols are wrongly synchronized. Due to the differential encoding, only two of them may be wrongly demodulated. This causes, however, a systematic error, which allows the system to operate only up to minimum BER. Therefore, in [106] it has been even observed that for continuous transmission, the system works better even without a Correlation Estimator, specially for high E_b/N_0 at the receiver.

New Proposed Solution: The interchange of the position of the Costas Loop and Polyphase Clock Sync blocks did not improve the BER performance. A new solution found to the described problem was the addition of a second Correlation Estimator block to the GNU Radio model, between the Polyphase Clock Sync and the Costas Loop, as it is shown in Fig. 7.3. In this way, the Costas Loop receives a tag that is placed in the proper position and does not suffer from the problem of $\frac{\pi}{2}$ - ambiguity anymore with this modification. Thus, the system is capable of working with a normal QPSK modulation instead of DQPSK.

7.3.4.3 Demodulation and Frame Decoding

The demodulation is carried out by a Costellation Decoder block. This has the same parameter configuration as the Costellation Modulator at the transmitter. After the Costellation Decoder, the GNU Radio model works at bit level.

The next block is the self-designed TDMA Frame Decoder It detects and interprets the frame, i.e. it regroups the frame bits into the components of the frame structure: preamble, address, and payload.

Frame Decoder without Trigger: The frame decoder must first find the frame preamble. In the first version of the frame decoder, as it was presented in [106] and [119], the detection was carried out only at bit level. The Correlate Access Code - Tag from GNU Radio was employed to achieve this. This last block compares the input bit stream bit with a predefined bit combination. If the Hamming distance between them is smaller than a configurable threshold, the preamble is detected. However, the Correlate Access Code - Tag block can correlate the received stream with a maximum of 64 bits, that is half the length of the preamble. The last 64 bits of the preamble are used for the correlation, since the symbols that correspond to first 64 bits are normally not perfectly synchronized, and the possibility of wrong decoded bits in this half is higher that in the second half.

Frame Decoder with Trigger: This is a second and enhanced version of the frame decoder and it is the one employed for the tests in this work. The TDMA Frame Decoder is triggered by the Correlator Estimator. That means, the frame decoder looks for the preamble in the current data stream only if the tag from the Correlator Estimator is found along this data stream. This increases the reliability of the system, due to redundancy. Besides, the correlation at the Correlator Estimator is performed at a symbol level, what it is more robust than a correlation at bit level[3]. The reason why the correlation at bit level is still used is that, even when

[3] The correlation at symbol level is more robust because the ratio between the maximum of the correlation and the highest side lobe is bigger than with a correlation at bit level. And additionally, because the correlation at symbol level employs the entire preamble, and at bit level only the second half.

employing two Correlator Estimator blocks, it was observed that there is still a possibility that the tag inserted by the second Correlator Estimator is placed in the wrong sample. The probability of such a constellation is very low but it would have a critical consequence in the frame decoding if there was no correlation at bit level: An offset of only one sample (which in this case equals one symbol) would produce a wrong decoding of a complete frame.

The preamble is removed once it is detected by the Correlate Access Code - Tag block. Then, the next byte, that corresponds to the address, is analyzed. It is compared to all available addresses, and the most similar, i.e. the one with the smallest Hamming distance, is chosen as received address. Finally, the TDMA Frame Decoder was designed to react to the frames received from a particular address, which is given as one of the block parameters. Then, if the received address is the same as the one set in the TDMA Frame Decoder, the address is removed from the frame and the remaining information, i.e. the payload, is transferred to the next block of the flow graph. If the addresses are not equal, the frame is discarded.

After the TDMA Frame Decoder, a Data Sink is placed. This block, the last one of the receiver flow graph, stores the received payload bits in a file for the offline post-processing.

7.4 Channel Coding

The aim of channel coding is to increase the quality of data transmission by error detection and correction. The cost of channel coding is a decrease in the net bit rate of information bits, since the amount of bits to be transmitted raises without increasing the received information bits after decoding. Furthermore, the cost of the implementation (hardware and/or software) also increases. Normally, there are two different approaches when implementing channel coding: Either the gross bit rate increases keeping the net bit rate constant or the gross bit rate remains constant and the net bit rate decreases. In this project, the second alternative was chosen, since the required gross bit rate of 2 Mbps was originally oversized to allow channel coding.

7.4.1 Selection of the Channel Coding Technique

There is extense available literature about channel coding techniques, as for example [44, 70, 72]. In the framework of this project, no new channel coding was developed. The suitability of two channel coding techniques for the wireless in-battery approach was investigated in [112]: BCH codes and convolutional codes. BCH codes were taken as a representative case of block codes. They were chosen among other analyzed block code techniques, due to their good compromise between performance and complexity. However, convolutional codes were found to be a more suitable alternative for the module-wise case of the in-battery communication in this work. They require a much lower complexity for the frame length under consideration. Therefore, convolutional code was the only alternative implemented with the rapid prototyping platforms.

7.4.2 Convolutional Codes

Convolutional codes are generated by passing a sequence of information bits through a LFSR. The LFSR consists of an input of k bits, a length of L stages (each of them of k bits), and n linear algebraic function generators [72]. A convolutional encoder generates an output sequence of n bits for each input sequence of k bits. In the framework of convolutional codes, L is denoted as constraint length. The code rate R is the ratio between information bits (bits at the input of the encoder) and transferred bits (bits at the output of the encoder). It is calculated as

$$R = \frac{n}{k} \tag{7.1}$$

for convolutional codes [70].

There are two ways of decoding convolutional codes: Hard or soft decision. The last one offers a considerable increase of performance in terms of BER, with a relatively small increase of complexity. This is an important advantage over block codes, since with this last coding method the implementation of soft decision decoding is considerably more complex than hard decision decoding. However, only hard decision was completely tested in the framework of this work, i.e. under both transmission modes and TDMA. More details are given below.

GNU Radio already offers the coder and decoder blocks for both hard and soft decision: FEC Extender Encoder and FEC Extended Decoder, as encoder and decoder, respectively. However, the available blocks only offer the possibility of working with a fixed constraint length $L = 7$ and a code rate $R = 1/2$. Therefore, a convolutional code with these parameters was employed for channel coding in order to shorten the time of development. This particular convolutional follows the so-called NASA standard.

7.4.2.1 Hard Decision Decoding

The implementation of hard decision decoding requires the following modifications respect to the model without channel coding:

- In the transmitter, the FEC Extender Encoder[4] block was inserted after the Vector Source that generates the address and payload, and before a Multiplexer, which adds the preamble, in order to complete the frame. That means that only the address and payload are encoded. Otherwise, the original preamble would lose its good ACF properties if it were encoded.
- In the receiver, the TDMA Frame Decoder was modified: First the frame synchronization is required (by means of the preamble) but then, since the address is encoded, it must be decoded before analysis. Therefore, the TDMA Frame Decoder was modified and divided into two blocks and the FEC Extended Decoder inserted between them:
  - Preamble Decoder: It synchronizes the frame by means of the preamble. At its output, the address and payload are present, but still encoded.
  - Access Code Decoder: It receives the address and payload uncoded from the FEC Extended Decoder. The address is checked and if it is the right one, the payload is placed at the block's output.

[4] FEC: Forward error correction.

All the results related to the use of channel coding in prototyping platforms under burst and TDMA transmission modes correspond to hard decision decoding.

7.4.2.2 Soft Decision Decoding

Both Costellation Decoder and Preamble Decoder blocks should operate with soft decision in order to perform soft decision decoding in a similar way as hard decision, i.e. placing the FEC Extended Decoder after the Preamble Decoder. The soft decision decoding is natively supported by the Costellation Decoder. However, the Preamble Decoder should have been adapted. A complete adaptation was not finished, due to time constrains. Nevertheless, two alternatives were tried, tested and presented in [112].

First Implemented Alternative of Soft Decision Decoding: The soft decision decoder was placed between the Costellation Decoder and the TDMA Frame Decoder. The output of the Costellation Decoder must be set to soft decision. The problem with this alternative appears when operating in burst transmission mode: The first and last bits of the frame are decoded with noise, so they are wrongly decoded. A solution to this problem was the addition of some pilot bits at the beginning and end of the frame, in order to bring (at the beginning of the frame) and leave (at the end of the frame) the Viterbi decoder in a known state. Nevertheless, these bits represent an energy lost, since they do not carry information. This produces a displacement of the BER vs E_b/N_0 curve to the right (to higher E_b/N_0 values), which is equivalent to a performance loss. The obtained performance was not clearly better than with hard decision decoding, or even worse for high E_b/N_0 values [112]. Therefore, this alternative was not longer considered.

Second Implemented Alternative of Soft Decision Decoding: Preamble Decoder was modified: The correlation at bit level was removed and the frame synchronization was realized only by means of the tag of the second Correlation Estimator. In this way, the Soft Decoder can be placed after the Preamble Decoder block. The advantage of this alternative is that the Soft Decoder receives frames after the frame synchronization, i.e. only bits from the frame's address and payload. It does not receive bits decoded from noise in the channel, when no signal is sent. Unlike the first alternative, the second one does not need any extra bits before and after the frame, which is the reason of the performance loss with the first alternative.

The problem with this second alternative was that the frame synchronization was not as robust as the original model without channel coding. The issue and its consequences have been explained in Section 7.3.4.3. The wrong position of the tag occurs, however, randomly and very sporadically: It was empirically observed that it occurs around once per hour. Since the duration of the measurements in continuous transmission mode is much shorter than with the other two modes, most of the measurements in continuous transmission mode produced a result according to the expectation: The improvement should be theoretically equivalent to a displacement of 2.2 dB to the left of the BER vs E_b/N_0 curves. The results depicted in Section 7.5.2 show the improvement, which are quite close to the theoretical expectation. However, the results were much worse than without coding in burst transmission modes, because the complete measurement takes some hours and some frames with a wrong position of the tag occurred.

Due to time constrains, this second alternative was not further improved. However, the results with soft decision decoding and continuous transmission mode are shown in this work and compared with the hard decision decoding for the same case.

7.5 Performance Evaluation of the Physical Layer

The performance of the communcation system was evaluated in terms of BER and percentage of packet loss as functions of the E_b/N_0 measured at the ADC of the receiver.

7.5.1 Method for Obtaining the BER Curves

E_b/N_0: To calculate the E_b/N_0, the SNR must be first obtained. The SNR is analytically calculated as

$$SNR = \frac{S}{N} = \frac{\mathrm{E}\left\{|s(t)|^2\right\}}{\mathrm{E}\left\{|n(t)|^2\right\}} = \frac{\sigma_S^2}{\sigma_N^2}. \tag{7.2}$$

However, both signal and noise power cannot be measured simultaneously within the same bandwidth. Therefore, the SNR must be estimated at the receiver. There are several approaches, but as in this case the channel under analysis is static, the SNR can be estimated by the following the next steps:

- The received signal power is measured at the receiver, while the transmitter does not send any signal. This corresponds to the noise power (σ_N^2).
- The received signal power is measured at the receiver, while the transmitter sends a signal. This corresponds to the addition of signal power (σ_S^2) and the noise power , i.e. $\sigma_S^2 + \sigma_N^2$.

Then, it can be assumed that the properties of noise are the same in both measurements when they are performed almost simultaneously, without changing the measurement conditions. That means, σ_N^2 is the same in both measurements. Additionally, as the bandwidths of the measurements are the same, the SNR can be estimated as

$$SNR = \frac{S}{N} = \frac{\sigma_S^2}{\sigma_N^2} = \frac{(\sigma_S^2 + \sigma_N^2) - \sigma_N^2}{\sigma_N^2}. \tag{7.3}$$

Then, the E_b/N_0 is calculated at the receiver from the estimation of the SNR. The relationship between them is [3]

$$\left.\frac{E_b}{N_0}\right|_{dB} = SNR|_{dB} + 10\log_{10}(n) - 10\log_{10}(sps), \tag{7.4}$$

where k is the amount of information bits per symbol. Both the modulation order and the code rate (in case that channel coding is employed) have an influence on this value: $k = \log_2 M \cdot R$ ($R = 1$ for uncoded information).

However, the E_b/N_0 is not constant along the different devices that compose the receiver chain. In fact, each device decreases the E_b/N_0 by introducing additional noise. The relationship between the input and output SNR at a device is known as noise figure (F) [13]. The values of signal obtained at software level (i.e. GNU Radio) are the digital values output by the USRP's ADC. The E_b/N_0 measured at this point is denoted as $(E_b/N_0)_{(ADC)}$ and it is lower than the E_b/N_0 at the SDR receiver input, which is denoted as $(E_b/N_0)_{(Rx)}$. The relationship between $(E_b/N_0)_{(ADC)}$ and $(E_b/N_0)_{(ADC)}$ is the noise figure:

$$F = \frac{(E_b/N_0)_{(Rx)}}{(E_b/N_0)_{(ADC)}}. \tag{7.5}$$

The $(E_b/N_0)_{(ADC)}$ is estimated with the values obtained at software level following 7.3 and 7.4. The value of the noise figure depends on the employed front end. In the case of the WBX daughter board the typical value is 5 dB according to the manufacturer [68]. Then, $(E_b/N_0)_{(Rx)}$ is obtained in dB as

$$(E_b/N_0)_{(Rx)}\,|_{dB} = (E_b/N_0)_{(ADC)}\,|_{dB} + 5\,dB. \tag{7.6}$$

In this work, the system performance will be analyzed as a function of the $(E_b/N_0)_{(ADC)}$, in order to evaluate the performance of the developed models for the transmitter and the receiver[5]. The reason is that the performance in terms of $(E_b/N_0)_{(Rx)}$ is highly dependent of the quality of the receiver chain hardware components, especially the LNA filter. These components vary according to the employed SDR platform or even to the selected daughter board in the case of the USRPs [68].

BER: It was already explained in Section 6.2.1. It is obtained by means of (6.8). The BER is calculated only over the bits from frames that are properly detected by the TDMA Frame Decoder at the receiver.

---

[5] In [106] the BER was also plotted as a function of $(E_b/N_0)_{(ADC)}$, even though it was not explicitly mentioned.

Packet Loss: It indicates the difference between the amount of transmitted and received packets. It is presented as a percentage of the transmitted packets. It should be ideally zero. A higher value means that some packets, i.e. frames, were not properly detected by the TDMA Frame Decoder at the receiver. The percentage of packet loss is calculated as

$$\text{Packet Loss [\%]} = \frac{\text{\#transmitted frames} - \text{\#received frames}}{\text{\#transmitted frames}} \cdot 100\%, \quad (7.7)$$

Obtaining the Desired $(E_\mathbf{b}/N_\mathbf{0})_{(\text{ADC})}$ Values: The performance of the system must be measured for relatively low $(E_\text{b}/N_0)_{(\text{ADC})}$ values, below 15 dB. Consequently, the signal power density at the receiver is in the same order of magnitude as the noise power density or only one order of magnitude bigger. The noise spectral density in the channel is very low, due to its thermal origin, as explained in Section 5.3. The addition of noise and received signal would still have a very small amplitude compared to the dynamic range of the receiver USRP's ADC, which is [−1 V, +1 V]. A similar problem will occur at the transmitter side. The consequence of operating with signal amplitudes significantly smaller than the dynamic range of the DAC and the ADC is that only the less significant bits of them are used. This produces a strong impact of the quantization noise and, hence, a performance loss.

Since the noise has been assumed to be AWGN, the solution was to generate own noise by means of an AWGN source in GNU Radio at the transmitter USRP and send it together with the desired signal. The power density of this self-generated noise is much higher than the channel noise, so the last one can be practically neglected. In this way, the power density and amplitude of the transmitted signal can be set to higher values still keeping the desired $(E_\text{b}/N_0)_{(\text{ADC})}$ range, but taking advantage of the entire dynamic range of the transmitter USRP's DAC. Also the signal at the receiver side is higher in this way, taking more advantage of the dynamic range of the receiver USRP's ADC. Since the channels between transmitter and receiver are practically flat within the used bandwidth, the noise at the receiver still has practically AWGN characteristics. If the channel would be strongly frequency-selective, then frequency shaping could be applied on the self-generated noise at the transmitter. That means, the CTF_BB between transmitter and receiver USRPs should be first measured (for example, employing the techniques developed in [103, 114]) and then, the inverse of this $\mathrm{CTF}_{\mathrm{BB}}$ should be applied to the spectrum of the self-generated noise.

Automation of the Measurement Process: The measurement process with the demonstrator consists in connecting two USRPs (one as transmitter and the other one as receiver) to the testbed and measuring the relationship between BER and the Packet Loss with $(E_{\mathrm{b}}/N_0)_{(\mathrm{ADC})}$. Long data streams must be sent between transmitter and receiver for different values of $(E_{\mathrm{b}}/N_0)_{(\mathrm{ADC})}$ [119]. The BER and Packet Loss must be measured for each of them.

The communication system models in GNU Radio were designed in a way that either each USRP can be managed by its own host computer or a unique host computer can be shared by all USRPs. A Bash script was written for a fast and easy automation of the measurement process, but this script requires all the USRPs in the demonstrator to share one host computer.

The first step of the automation of the measurement process is to estimate the $(E_{\mathrm{b}}/N_0)_{(\mathrm{ADC})}$ at the receiver for a known amplitude of noise and a known amplitude of transmitted signal. Then, the amplitude of the signal is kept constant and the necessary self-generated noise amplitude values for the desired $(E_{\mathrm{b}}/N_0)_{(\mathrm{ADC})}$ values are calculated. Then, the following process is repeated for each different obtained noise amplitude value: The desired signal and noise are sent together from transmitter to receiver. The BER and Packet Loss are calculated for each $(E_{\mathrm{b}}/N_0)_{(\mathrm{ADC})}$ value.

Estimation of the Transmission Power: Introducing (6.10) in (6.11), P_{Tx} can be also calculated as

$$P_{\mathrm{Tx}} = \frac{(E_{\mathrm{b}}/N_0)_{(\mathrm{Rx})} \cdot N_0 \cdot \frac{f_{\mathrm{s}}}{sps}}{\frac{1}{K^2} \cdot \sum_{k=0}^{k=K-1} \left| H_{\mathrm{TxF}}[k] \cdot H_{\mathrm{ch}(i,l)}[k] \right|^2}. \tag{7.8}$$

The value of the channel attenuation from the transmitter output up to the receiver input, i.e. the factor $\frac{1}{K^2} \cdot \sum_{k=0}^{k=K-1} \left| H_{\mathrm{TxF}}[k] \cdot H_{\mathrm{ch}(i,l)}[k] \right|^2$, was calculated for each of the four slaves at the center frequency and converted in decibel, obtaining the following values: –25.9 dB, –19.3 dB, –23.1 dB, and −21.3 dB for the channels M-S1, M-S2, M-S3, M-S4, respectively. In order to estimate the required transmission power, the smallest of this values (–25.9 dB) is employed in (7.8). N_0 is the noise spectral density in the channel, i.e. at the receiver input. It was already estimated in Section 5.3 and it is equal to −173.5 dBm/Hz.

The required transmission power must be calculated for a BER = 10^{-3}. Hence, the corresponding value of $(E_b/N_0)_{(\mathrm{ADC})}$ is obtained from the BER vs. $(E_b/N_0)_{(\mathrm{ADC})}$ curves and converted into $(E_b/N_0)_{(\mathrm{Rx})}$ by means of (7.5). Finally, the value of $(E_b/N_0)_{(\mathrm{Rx})}$ in linear scale (not in dB) is introduced in (7.8) to calculate the required transmission power.

7.5.2 Results and Discussion

7.5.2.1 Continuous Transmission Mode

Fig. 7.4 shows the results of BER cuves and packet loss for the simulation and measurement of the continuos transmission mode, both with and without channel coding. Additionally, they are also compared to the theoretical curve of QPSK under an ideal AWGN channel.

Without Channel Coding: The results of the BER curves of the simulated model under the baseband equivalent of the emulator channel and the theoretical case are very similar, even when the simulation includes a frequency offset between transmitter and receiver oscillators of 30 Hz. This proves that the baseband equivalent channel has an almost flat frequency response and also the good functioning of the simulation model.

The measured results with the USRPs are better than the ones previously published for the first version of this model [106, 119], especially in the case of continuos mode with Correlation Estimator. The main reason for this improvement is the addition of the second Correlation Estimator. This solved the systematic error caused by the wrong position of the tag after the Polyphase Sync Clock. The measured BER curve with the USRPs has an offset of around 2.4 dB with respect to the simulated curve. However, this result is also still much better that the one published in [106, 119]. The necessary $(E_b/N_0)_{(\mathrm{ADC})}$ to fulfill the requirement of BER $= 10^{-3}$ is 6.8 dB in the simulation and 9.2 dB in the measurement. Employing these values in (7.8), a transmission power of –75.79 dBm and –73.49 dBm are obtained for the simulation and the measurement, respectively. It must be remarked that both in simulation and measurement, no horizontal asymptote appears in the curves in the considered $(E_\mathrm{b}/N_0)_{(\mathrm{ADC})}$ range, as it was observed in [106, 119].

Using Hard Decision Channel Coding: The simulated and measured BER curves of the continuous transmission mode with channel coding have an offset of around 2.2 dB, similar to the case without channel coding. The BER at $(E_\mathrm{b}/N_0)_{(\mathrm{ADC})} = 6.8\,\mathrm{dB}$ in the simulated case is 6.6×10^{-6}, i.e. the BER has been decreased by a factor of 151.5. However, in the case of the measurement result, the BER at $(E_\mathrm{b}/N_0)_{(\mathrm{ADC})} = 9.2\,\mathrm{dB}$ is 2.4×10^{-7}, which means an improvement of a factor 4167. As well as without channel coding, no horizontal asymptote appears in the considered $(E_\mathrm{b}/N_0)_{(\mathrm{ADC})}$ range.

The packet loss is bigger with channel coding than without. This occurs not only in continuous transmission mode, but with all the transmission modes under test. The reason is the following: For the the same $(E_\mathrm{b}/N_0)_{(\mathrm{ADC})}$ value, the SNR is with channel coding lower than without, due to the displacement of the curve of $10 \cdot \log R = 3\,\mathrm{dB}$. Since only the preamble is employed for the frame detection and it is not coded, the preamble suffers from a higher noise power density (if the same transmit signal power density is employed) for the same $(E_\mathrm{b}/N_0)_{(\mathrm{ADC})}$ when channel coding is used. The amount of erroneously detected preambles increases and consequently, the packet loss too.

Using Soft Decision Channel Coding: The comparison between hard and soft decision channel coding is depicted in Fig. 7.5. The better performance of the soft decision decoding is clearly visible. For example, according to the measurements, soft decision decoding offers a margin gain[6] of 2.3 dB respect to hard decision decoding at $\mathrm{BER} = 10^{-6}$. That is very close to the value of 2.2 dB that would be theoretically obtained for a $\mathrm{BER} = 10^{-6}$ in an AWGN channel [112].

---

6 Margin gain is the difference in the $(E_\mathrm{b}/N_0)_{(\mathrm{ADC})}$ values at which both alternatives reach a particular BER.

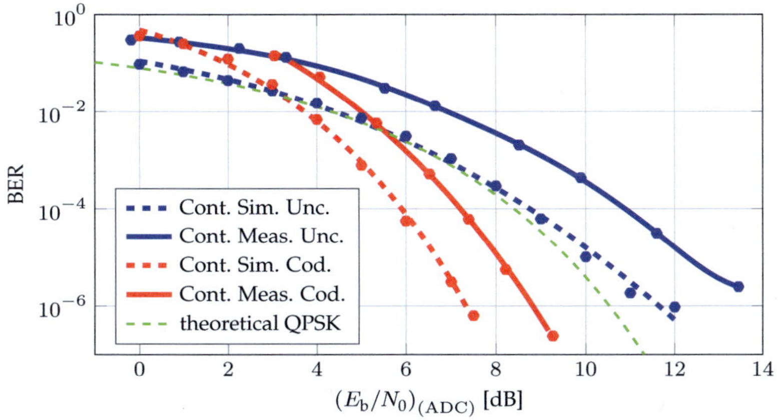

(a) BER

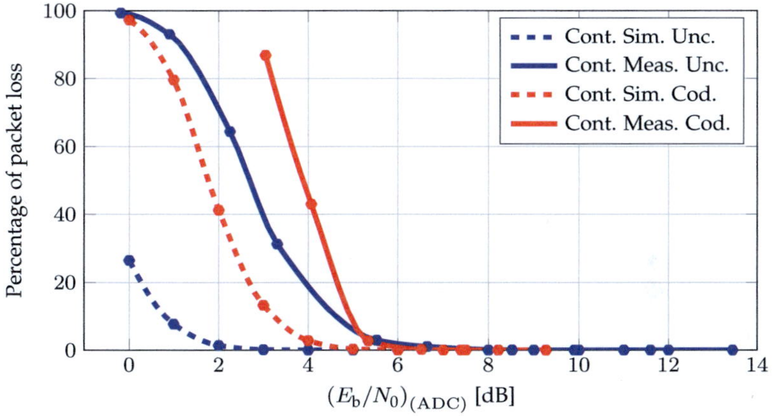

(b) Packet Loss

Figure 7.4: Performance of the communication system prototype in simulation and measurements within the small channel emulator, in continuous transmission mode and with helix antennas

The packet loss rate is smaller with soft decision at a low $(E_b/N_0)_{(ADC)}$. The cause is the frame detection at bit level in the hard decoding, which is less robust than the frame detection at symbol level. Therefore, the frame decoder may be no able to detect preambles at a low $(E_b/N_0)_{(ADC)}$.

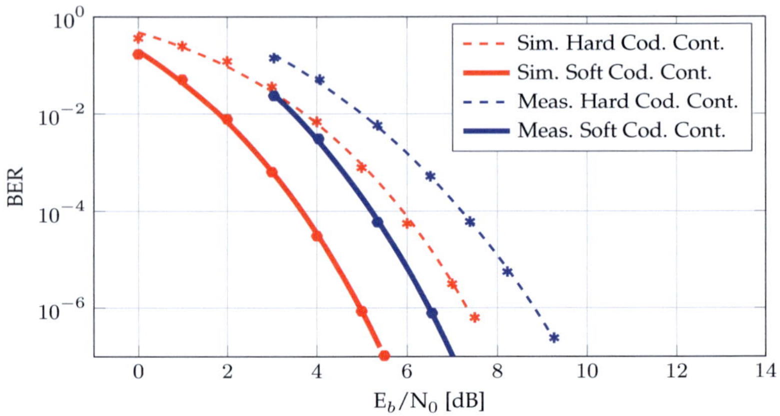

(a) BER

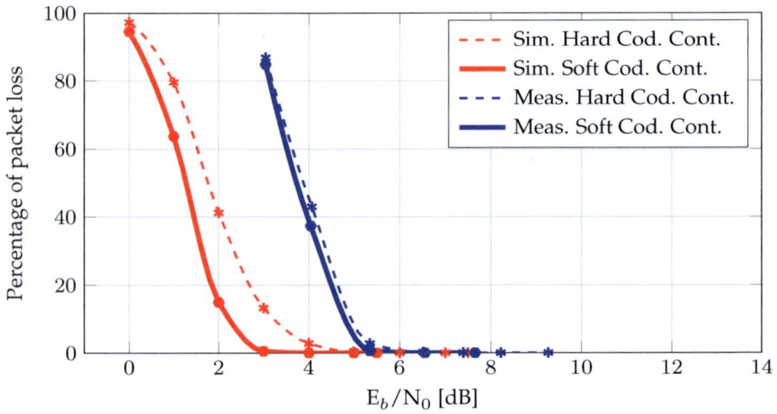

(b) Packet Loss

Figure 7.5: Performance comparison between hard and soft decision channel decoding with continuous transmission mode

7.5.2.2 Burst Transmission Mode

The results of BER cuves and packet loss rate for the simulation and measurement of the burst transmission mode, with and without channel coding, are depicted in Fig. 7.6.

Without Channel Coding: The obtained BER curves with burst transmission mode for both simulation and measurement are very similar to the ones of the continuous transmission mode. They are almost coincident for $E_b/N_0 < 10\,\text{dB}$. The BER curves of the burst transmission mode are better for higher values of $(E_b/N_0)_{(\text{ADC})}$. This is a big difference in comparison with the results published in [106] obtained with the first version of this implementation. With that model, the BER curves of the burst transmission mode were always worse than with continuous transmission mode, both in simulation and measurement. The reason is that the improvement in the TDMA Frame Decoder had a higher impact on the burst transmission mode than on continuous transmission mode. There is always a transmission pause between frames in burst mode, whose temporal length is much bigger than the one of a frame. Consequently, there is a possibility that the noise received during this pause is interpreted as a frame by the Frame Decoder. If this occurs, a complete frame is wrongly decoded. The results obtained in this work prove that the new symbol level triggered version of the TDMA Frame Decoder is more robust against the wrong detection of frames when there is only noise in the channel.

The necessary $(E_b/N_0)_{(\text{ADC})}$ to fulfill the requirement of BER $< 10^{-3}$ is 6.8 dB in the simulation and 9.2 dB in the measurement, same results as with continuous transmission mode. Therefore, the required power transmission values are also the same as with continuous transmission mode in both cases.

Using Hard Decision Channel Coding: The BER curves are not overlapped with the ones obtained with continuos transmission mode, but they still offer an important performance improvement against the results of the uncoded model. The BER in the simulation at $E_b/N_0 = 6.8\,\text{dB}$ is 6.3×10^{-3}, i.e. the BER is improved a factor 159. In the measurement, a BER of 2.45×10^{-3} is observed at $E_b/N_0 = 9.2\,\text{dB}$, resulting in a BER 4082 times better than without channel coding.

7.6 Implementation of the MAC Layer

A block diagram of the MAC layer implementation is shown in Fig. 7.7. Many of the blocks presented in this diagram include other blocks that have already been explained in the implementation of the physical layer. As already explained in Section 6.3, a TDMA protocol is employed.

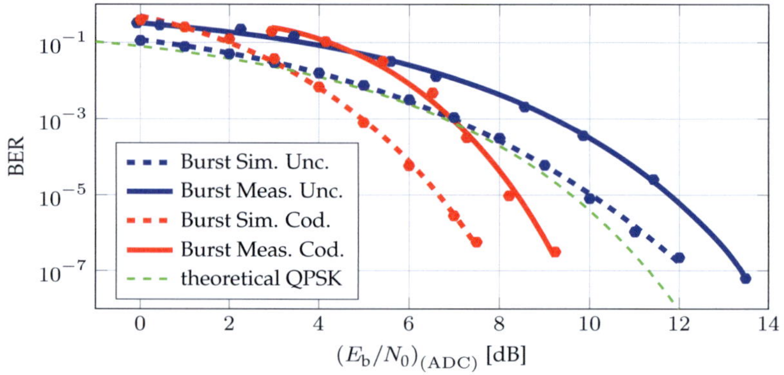

(a) BER

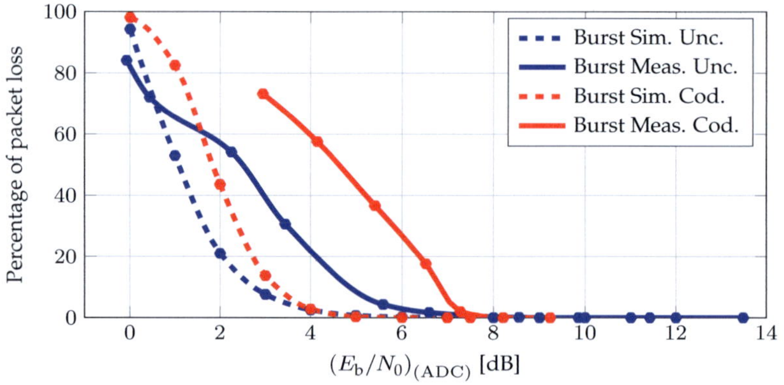

(b) Packet Loss

Figure 7.6: Performance of the communication system prototype in simulation and measurements within the small channel emulator, in burst transmission mode and with helix antennas

In the implementation of the physical layer, each communication node was either transmitter or receiver. In the MAC layer implementation, each communication node is a transceiver, i.e. it includes a transmitter and a receiver. As the mode of operation is half-duplex, each communication node needs only one antenna and both the transmitter and the receiver operate in the same frequency band, but at different times. The UHD's tags can be employed to easily and safely change the operation mode of the USRP between transmitter and receiver, as explained in the implementation of the Burster in Section 7.3.2.3.

7.6.1 TDMA Protocol

Every communication sub-cycle begins with a master broadcast frame. Therefore, the master begins the sub-cycle operating as transmitter. The slaves start the sub-cycle as receivers, waiting for the master frame. When this frame is detected, the slaves know at which instant of time the sub-cycle has begun. Every slave will configure itself to change its operation mode to transmitter and send its frame in the assigned time slot. Once a slave finishes sending its frame, it will operate as receiver again, waiting for the next master frame. On the other side, the master starts to operate as receiver once its frame is sent. The master stays in this mode, receiving and decoding the frames coming from the slaves, until it is time to send a new master frame to start a new sub-cycle. The described process is repeated continuously. The structure of the transmitters and receivers of both master and slave are described below.

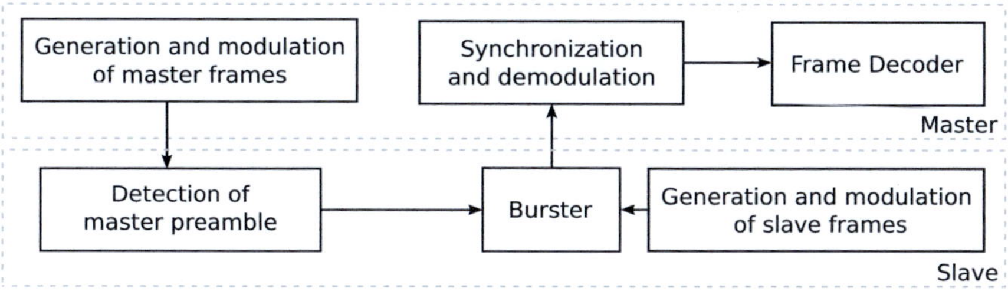

Figure 7.7: Block diagram of the MAC layer implementation

7.6.2 Master

The structure of the master's transmitter is the same as the one explained in Section 7.3.2 for the transmitter of the physical layer implementation. The only difference is the structure of the frame employed by the master during the broadcast signal at the beginning of the sub-cycle. This frame does not need to transmit any particular information to the slaves, it must be only detected by them. Therefore, only a preamble is sent as master frame. This preamble is a Gold code of 64 symbols. By means of the UHD's tags generated by the master's Burster (the TDMA Frame Decoder block), the USRP changes to transmitter mode and it sends the frame master. Then, the USRP changes automatically to receiver mode, waiting for the data frames coming from the slaves.

The master's receiver has the same structure as the one described in the receiver of the physical layer. With the current design of the TDMA Frame Decoder block, there must be one these blocks for each slave in the master's receiver, since each TDMA Frame Decoder is associated to only one slave, according to the content of its address field.

7.6.3 Slaves

The slave's receiver employs a Correlation Estimator block for the detection of the master frame. When this frame is recognized, i.e. when the correlation between the received signal and the sent preamble exceeds a certain threshold, the Correlation Estimator propagates a tag, which is received by the next block, the Slave Frame Tagger. This last block stores the time at which the tag is received as starting instant of the sub-cycle. Each slave is previously assigned with a certain time slot, which is provided by the Slave Frame Tagger block. This time slot is the offset with respect to the beginning of the sub-cycle, so the addition of both times is used by the Slave Frame Tagger to tag the beginning of the frame (burst) with a tx_time tag, indicating the time at which the frame is sent. The tx_sob and tx_eob are also added by the Slave Frame Tagger.

The structure of the slave frame is the same as the one explained for the physical layer in Section 7.3.2.1. The structures of both master and slave frames are represented Fig. 7.8.

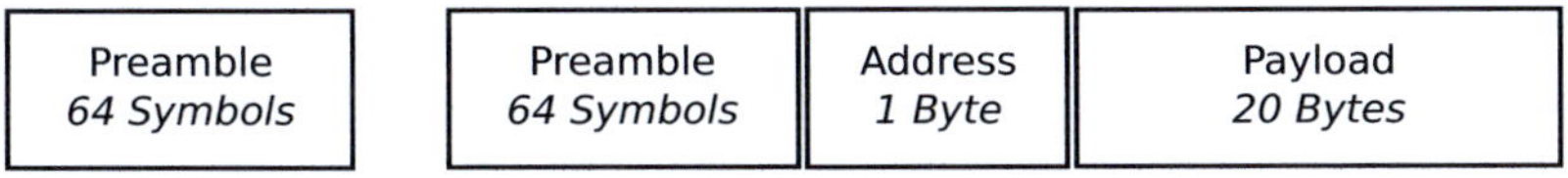

(a) Master frame **(b)** Slave frame

Figure 7.8: Structures of the master and slave frames

7.7 Performance Evaluation of the Physical and MAC Layers

The BER is calculated as the ratio between the error bits received at the master and the transmitted bits by the slave. The packet loss is the ratio between not detected frames at the master (coming from the slaves) and the product of the amount of sent frames by the master and the amount of slaves. That means, there are two possible reasons for a lost packet: The first one is that the slave does not recognize the master frame and consequently, it cannot realize that a sub-cycle began and it does not send any packet. The second one is that the master's frame decoder did not detect the frame sent by the slave.

7.7.1 Method

The elements of the demonstrator and their connection is exactly the same as for the measurements of the continuous and burst transmission modes, but employing the GNU Radio models of the TDMA implementation. A similar bash script to the one explained in Section 7.5.1 was written for the measurement process automation.

7.7.2 Results and Discussion

Since the current model of the TDMA system cannot be simulated in GNU Radio without the USRPs, the BER curves of TDMA are compared in Fig. 7.9(a) against the measured curves of the burst transmission model, because the physical layer operates similarly in both cases. It can be seen in Fig. 7.9(a) that the curves without channel coding practically overlap up to an $E_b/N_0 = 11.5\,\text{dB}$. In the case of the curves with channel coding, TDMA performs even better for $E_b/N_0 < 9\,\text{dB}$. Then, the curves are again practically coincident. This means, that a TDMA system also performs much better than the previous version of the model published in [106], where its BER curve was considerably worse than the ones of the burst and continuous transmission modes. The reasons for this improvement in performance are the same as the ones explained for the burst transmission mode in Section 7.5.2. With the current models of the communication

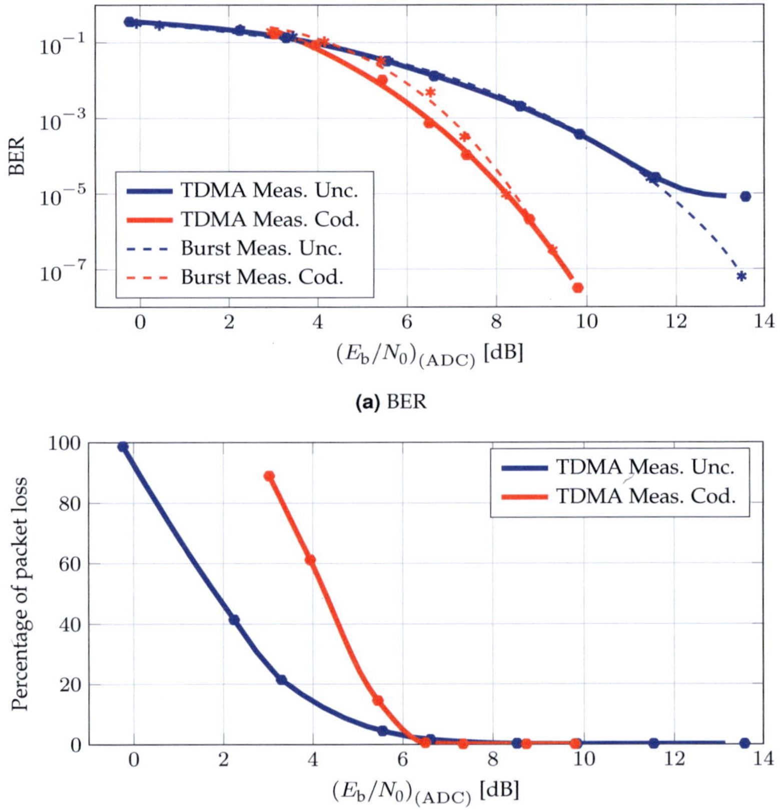

(a) BER

(b) Packet Loss

Figure 7.9: Performance of the communication system prototype in simulation and measurements within the small channel emulator, in burst transmission mode and with helix antennas system, the performance in terms of BER is practically the same with only the physical layer and with both physical and MAC layer.

The model without channel coding requires an $E_b/N_0 = 9.1\,\text{dB}$ for a BER $= 10^{-3}$. Then, applying (7.8), the required transmission power is –73.49 dBm. The model with channel coding has a BER $= 4.65 \times 10^{-7}$ for the same $(E_b/N_0)_{(\text{ADC})}$, i.e. the performance increases by in a factor 4650.

Measurement of the Propagated Signals with an Oscilloscope: The signals propagating inside the demonstrator were measured with an oscilloscope and they are shown in Fig. 7.10. One USRP as master and two as slaves were employed in this measurement. Two sub-cycles can be distinguished in the main window of the oscilloscope's measurement. The setup configuration in GNU Radio for the TDMA protocol was the following:

- Sub-cycle interval = 10 ms
- Time slot of the first slave = 2 ms
- Time slot of the second master = 4 ms

These values can be visually validated in the measurement. The first burst starting from the left is the master frame, and the next two are the responses of the slaves. The sub-cycle is repeated within the next three bursts.

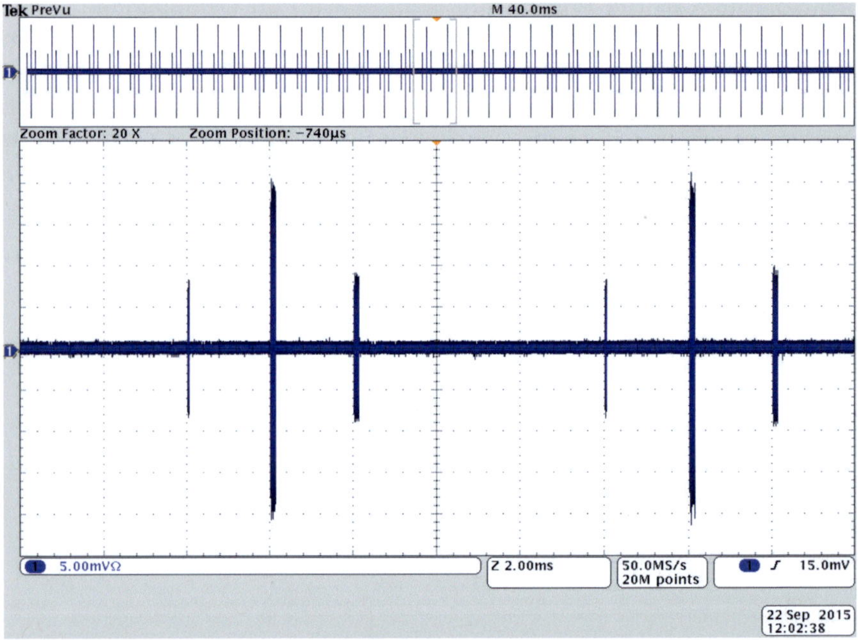

Figure 7.10: Screen dump of the oscilloscope measurement in demonstrator with two USRPs as slaves and running under TDMA transmission mode

8 Measurements in a Cold Automotive Battery

Near the end of the IntLiIon project, the access to a real automotive battery for experimentation was provided by one of the project partners. The employed off-the-shelf traction battery was the one used as reference for the small emulator. Two different types of measurements were carried out: First, measurements of S-parameters with a VNA, similar to the ones presented in Sections 3.4.3 and 4.3.2. Then, the performance of the communication system prototype presented in Chapter 7 was also measured in the battery. In both cases, the measurements were performed in a cold battery. No vehicle or any kind of load were connected to the battery.

Two major issues must be pointed out: The first one is that only a reduced amount of measurements could be carried out, due to the limited availability of the laboratory where the measurements in the battery were conducted. The second issue is that no CAD file with the detailed characteristics and physical dimensions of the battery was provided, due to the confidentiality of this information. Because of this, it was not possible to design a simulation model of the battery for electromagnetic simulations with the software CST MWS. Therefore, a validation or comparison of measured results in the battery against simulations was not possible. The results obtained are, however, presented and discussed in this chapter.

8.1 CTF Measurements

8.1.1 Setup and Method

The battery employed for the measurements, as well as the complete testbed, are depicted in the photos in Fig. 8.1 and in Fig. 8.2. Four holes were drilled at the upper side of the battery housing for mounting the antennas. The SMA connectors of the antennas were connected to coaxial cables through these holes. The antennas were connected to a VNA or to USRPs by means of these cables. Copper band was employed in order to fix the antenna connectors to the battery housing. This copper band also assures a good contact between the antenna ground and the battery housing. As explained in Section 2.3, this ensures a stable and accurate measurement.

In the design of this wireless approach proposed in Section 1.3 and shown in Fig. 1.2, the antennas should be placed above the cells, between their terminals. This was not possible in the battery employed for measurements in this project. This battery has metallic rectangular pipelines above each module and between the terminals, as it can be seen in Fig. 8.3(a). These pipelines are employed to cover and protect the communication wiring and the electronic boards that belong to the sensor system installed above each battery cell. If the proposed wireless communication system were employed, no communication wiring should be placed inside this pipeline. The pipeline could be then smaller, leaving space for the antennas above it. Another possibility would be to integrate

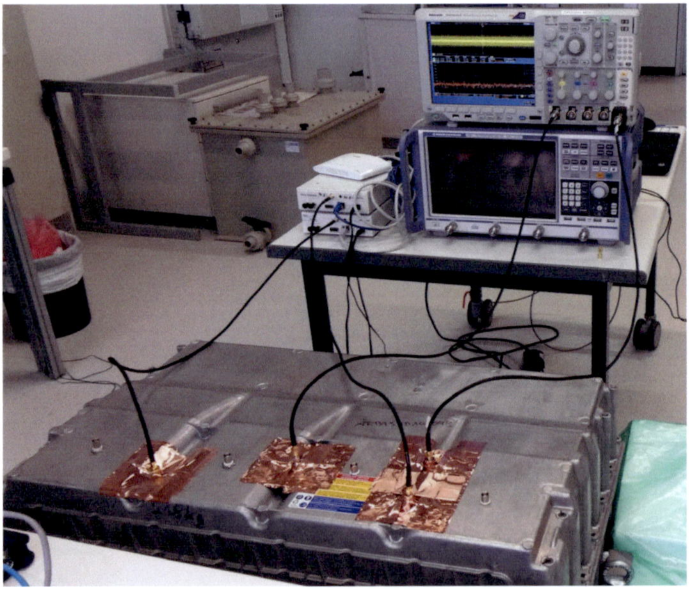

Figure 8.1: Setup for the measurements in an automotive battery

Figure 8.2: Setup for the measurements in an automotive battery, view from the front of the battery

the antennas into the electronic boards and to employ a cover made of plastic to protect the boards but without shielding the antennas.

The antennas could be placed only at the locations where no cell terminal connectors for the measurements in this battery were present. The customized helix antennas were employed for measurements in the first frequency sub-range, and the self-customized PIFAs for the third frequency sub-range. Only four positions were evaluated, due to time constrains. The position and mounting of the antennas can be seen in Fig. 8.2. and Fig. 8.3.

The cover at the upper side of the battery housing is not completely flat as the channel emulator used in this work. The battery housing cover has two regions which are almost parallel to the battery floor. But the housing top cover has a slope in the middle. Hence, the antennas were placed only at the side of the battery with the biggest free available space.

The length and width of the battery are practically the same as the ones of the small channel emulator, because this battery was taken as reference for the small emulator dimensions. However, the available height between the upper side of the cells and the housing top is not constant, in contrast to our emulator, due to the cover profile. The largest available height is between 3 cm and 4 cm, which is smaller than the emulator height.

(a) PIFAs

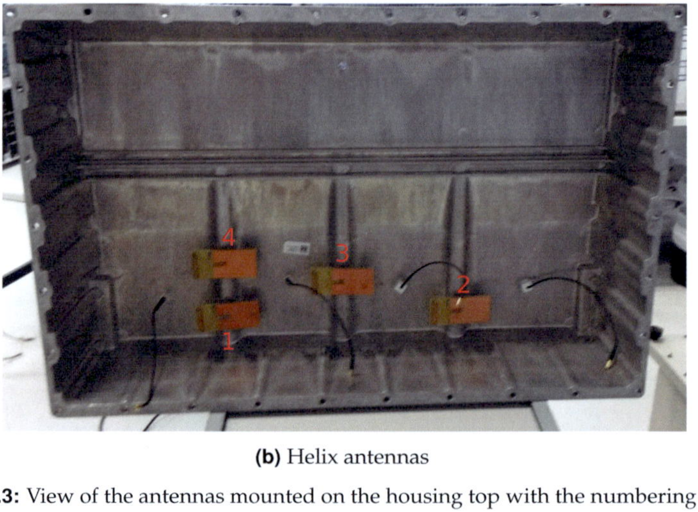

(b) Helix antennas

Figure 8.3: View of the antennas mounted on the housing top with the numbering of the antennas

8.1.2 Results

The numbers used as reference names of each antenna position are depicted in Fig. 8.3(b). The antenna in the position 3 played the role of the master for the measurements: The CTF between the antenna in this position and the remaining ones are the S-parameters shown in Fig. 8.4 for the helix antennas and in Fig. 8.5 for the PIFAs. Additionally, the similarity factors for the same cases were calculated and are presented in Table 8.1 taking S_{13} as reference case.

The CTFs observed in the measurements with the customized helix antennas at lower frequencies in the battery show a similar pattern as the ones measured in the emulator shown in Fig. 7.2. One important difference is that the CTFs in the battery have a much higher amount of resonant and notch frequencies, but most of them do not produce a high variation of the CTF. The reason is probably the coupling between the wire of the helix antenna and the environment components (housing, cells, etc.), as well as reflections on these elements.

The CTFs in the battery present a narrow region of higher attenuation around 350 MHz. There is even a clear and strong notch in the case of S_{23}. This issue is, however, also observable with this antenna design in Fig. 7.2. The similarity factor values are above 70% and it is visually clear that the three CTFs shown in 8.4 have similar patterns. Furthermore, the similarity factor values are higher than at higher frequencies, proving again the lower position dependence at lower frequencies.

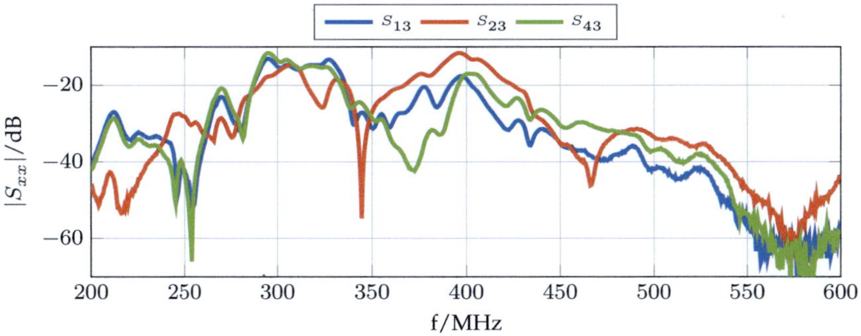

Figure 8.4: CTF measurements between helix antennas in a cold automotive battery

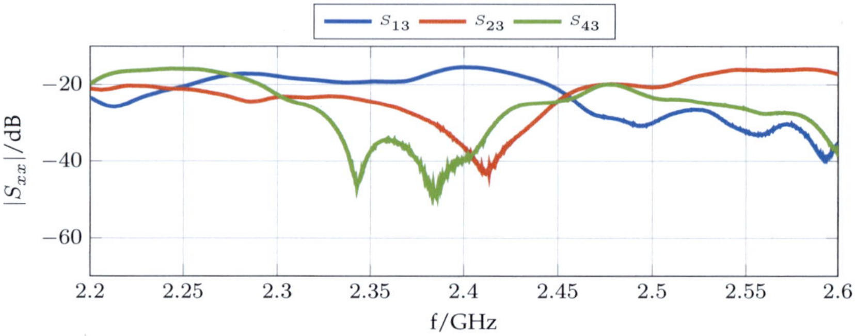

Figure 8.5: CTF measurements between PIFAs in a cold automotive battery

Table 8.1: Similarity factor between the reference channel S_{13} and the remaining channels S_{23} and S_{43}

S_{xx}	Similarity Factor with S_{13}	
	Helix (200 MHz – 600 MHz)	PIFA (2200 MHz – 2600 MHz)
S_{23}	72.7%	65.2%
S_{43}	83.7%	61.0%

In the case of the higher frequency range, the CTFs present less oscillations than at lower frequencies. The CTFs measured in the battery are less frequency-selective as the ones measured in the emulator. The CTFs in the battery show larger regions with a small variation of attenuation and the notches are not as deep as in the emulator.

8.2 Characterization of the in-Battery Channel

The channels measured in the battery were characterized in the same way as the simulated channels in Section 5.5. The channels under analysis were characterized with a bandwidth of 400 MHz, for the frequency interval between 200 MHz and 600 MHz with helix antennas, and between 2200 MHz and 2600 MHz with PIFAs. The results are shown in Table 8.2.

Two important conclusions can be drawn. The first one is that the coherence bandwidth of the channel inside the battery is also always larger than the symbol bandwidth of the communication system prototype used in Chapter 7. Thus, the communication system prototype still fulfills the requirements for operation in the battery. The second conclusion is that the coherence bandwidth obtained from the measurements in the battery is considerably smaller than the one obtained in the simulations of the emulator. The reasons are that the channels are not almost flat between 300 MHz and 400 MHz as in the simulations and the oscillations of the measured CTF. On the other hand, the coherence bandwidth with PIFAs is higher in the measurements in the battery than in the simulations, due to the large frequency ranges with almost constant attenuation.

Table 8.2: Results of the coherence bandwidth obtained for the channels measured in a battery in the frequency range between 200 MHz and 600 MHz with helix antennas, and between 2200 MHz and 2600 MHz with PIFAs

B_{coh} [MHz]		
Value	Helix	PIFA
Min	73.28	129.79
Mean $\pm$ Std	84.06 ± 12.82	204.28 ± 94.05

8.3 Performance of the Prototype in a Battery

The BER and packet loss rate of the communication system prototype were measured in the battery. However, only continuous transmission mode was investigated. The measurements of burst transmission method and the TDMA protocol would have taken some hours and there was no availability of the laboratory with the battery over such a time span.

The measurement with the USRPs at the battery was carried out with helix antennas in the first frequency sub-range, as in the case of the emulator. A value of 305.5 MHz was obtained as optimal center frequency and employed as center frequency. The remaining configuration of the system was exactly the same as in the measurements in Section 7.5.2.1 without channel coding.

The results appear in Fig. 8.6 compared to the measurements with USRPs in the small emulator. The results are practically coincident, because also in the battery the passband channel presents flat fading and does not introduce ISI. The minimum necessary E_b/N_0 to fulfill the requirement of BER $< 10^{-3}$ is 8.9 dB.

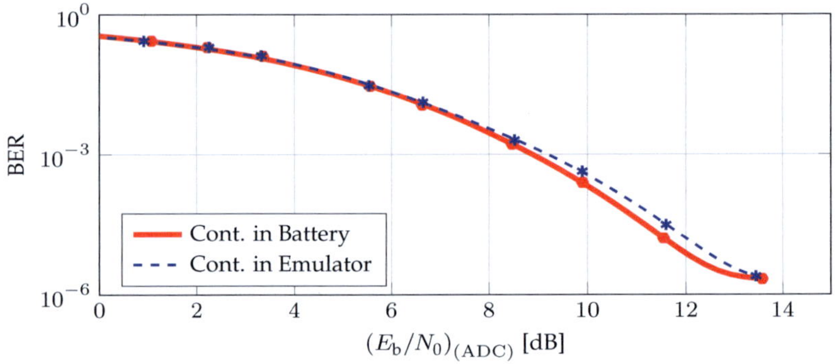

(a) BER

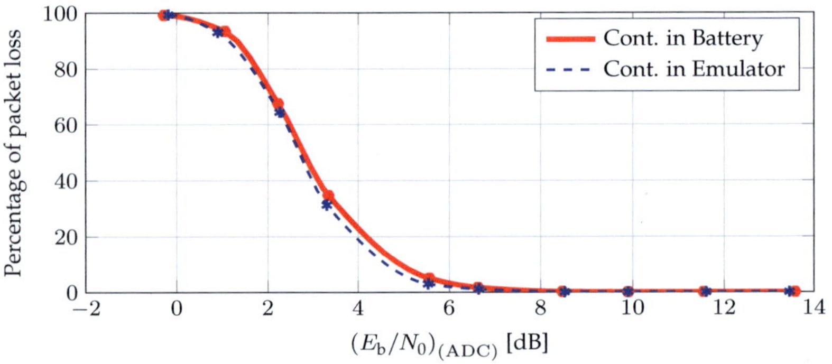

(b) Packet Loss

Figure 8.6: Performance of the communication system prototype measured in an automotive battery, in continuous transmission mode and with helix antennas

9 Conclusions and Future Work

The thesis was conducted under the framework of the IntLiIon project. This industrial research project was launched with the aim of developing novel communications techniques for the BMS deployed in traction batteries of electric and hybrid vehicles. The particular aim of this thesis was the investigation of a wireless data transmission approach for the management system of automotive batteries.

9.1 Summary

The following aspects were analyzed in the course of this doctoral thesis:

Measurement Testbed of the Wireless In-Battery Channel

Carrying out all channel measurements and performance evaluations of the communication modem prototypes in a real battery is not an easy task. It requires a large preparation. Batteries are also very expensive and difficult to handle, due to their dimensions, weight and risk of explosion. Therefore, a simple physical emulator of the wireless channel within automotive batteries was conceived based on the characteristics of this particular channel and presented in Chapter 2. Two emulators of different sizes were built, whose dimensions are based on commercial batteries. Antennas were placed inside them, in order to develop a complete testbed that can be used for several types of measurements.

Channel Measurements and Selection of the Antennas

In Chapter 3, several types of antennas were measured and compared over a frequency range from 100 MHz up to 5.9 GHz in both channel emulators. Based on the measurement results and on some issues related to the functioning of the testbed for frequencies above approximately 2.8 GHz, two antennas for two different frequency sub-ranges were chosen: Helix antennas for the lowest investigated frequency sub-range (between

100 MHz and 700 MHz) and PIFAs for the third investigated frequency sub-range (between 2100 MHz and 2700 MHz).

3D Electromagnetic Simulation of the Channel

Chapter 4 introduced the 3D electromagnetic simulation models of the measurement testbed that were developed for the software CST MWS. The accuracy of these models was validated against measurements with a VNA in three different situations, including the testbed.

Analysis of the Channel: Parametrization and Characterization

Chapter 5 dealt with a more detailed analysis of the channel by means of simulations of the channel emulator. First, the influence of different system parameters (such as emulator dimensions, antenna positions, amount of antennas, etc.) on the CTF between antennas was analyzed for the lower frequencies with helix antennas, as well as for the higher frequencies with the PIFAs. It was observed that the channel at low frequencies[1] and with a reduced number of antennas has the advantage of a very low dependence on the emulator dimensions. Additionally, there is a very low position dependence on the CTF form with a reduced number of antennas inside the emulator. However, the channel attenuation is strongly dependent on the distance between the antennas. This variation of attenuation vs. distance is even higher than at higher frequencies. This is an important handicap of the channel at low frequencies, since the transmission power must be calculated for the worst case.

On the other hand, the channels at higher frequencies are much more frequency-selective. Nevertheless, they have a positive characteristic: Many resonant frequencies of the different channels are coincident in each particular architecture.

The modeling of the baseband equivalent model of the channel and the noise in the emulator was also explained in Chapter 5. These models are necessary for software-based simulations of the communication system. Next, an algorithm for the selection of the optimal center frequency of the

[1] As explained in Chapter 3, the terms "'low" and "high" frequencies are not absolute values, but they depend on the relationship between the battery or emulator dimensions and the wavelength of the frequencies under consideration.

communication network was developed and employed in several simulated architectures. Finally, the communication channel was characterized by means of the coherence bandwidth for each of these architectures.

Design and Performance Simulation of the Physical and MAC Layers

In Chapter 6, the architecture of the modems (transmitter and receiver) for the physical layer was designed based on the results of the channel characterization and project specifications. The modems and the channel were modeled in Simulink and their performance was evaluated by simulations under the different architectures obtained from Chapter 5. The results show that less transmission power is necessary in order to achieve the same performance (BER) at higher frequencies. The characteristics of the MAC layer, which employs TDMA as channel access method and a master–slave hierarchy, were also described in Chapter 6.

Prototyping of the Communication System

Prototypes of the communication modems were developed in Chapter 7 by means of SDR platforms. The off-the-shelf USRPs were employed as SDR platforms and they were programmed by the software toolkit GNU Radio. In addition to the architecture proposed for the physical layer in Chapter 6, channel coding and TDMA protocol for the MAC layer were also implemented.

The performance of the communication system was measured in terms of BER and packet loss rate in the testbed based on the small channel emulator. The measurement results are slightly worse than the ones from simulations. However, it is possible to reach the required performance at values of E_b/N_0 always below 10 dB even in measurements, which implies very small values of transmission power.

Measurements in a Cold Automotive Battery

Chapter 8 presented the measurements that were performed in a real cold automotive battery. First, the CTF between four antennas was measured inside the battery, in a similar way as the measurements in the emulator-based testbed employed in previous chapters. This operation was carried out with both helix antennas and PIFAs. The channel was characterized based on these measurement results. Most of the conclusions drawn from these measurements and characterizations were similar to the ones drawn from the measurements in the emulator-based testbed, demonstrating the efficiency of the proposed method. Finally, the performance of the modem prototypes from Chapter 7 was also tested in the battery, yielding similar results as the ones in the emulator.

9.2 Conclusions

The most relevant results and conclusions of this work are the following:

- Among the considered types, the antennas with the best performance were helix antennas in the first frequency sub-range and PIFAs in the third and fourth frequency sub-ranges.
- Working at low frequencies could be considered as a good alternative for a small number of antennas, e.g. the case of a module-wise communication with a reduced number of modules. The characteristics of the communication channel enable the design of a communication system without requiring the previous knowledge of the CTFs between the master and every slave. This was called a *plug-and-play* system. But for a large amount of antennas, i.e. if the number of modules increases or in the case of a cell-wise communication, working at higher frequencies is a better approach. At higher frequencies, less transmission power is required and the antennas are cheaper and smaller.
- The communication system does not suffer from ISI in the evaluated channels for the required bit rate of 2 Mbps, even when BPSK or QPSK are used as modulation technique.
- The required BER can be achieved with a very small transmission power. This was proved with simulations and measurements with the modem prototypes. In the simulations in Section 6.2, the worst obtained value was −64.41 dBm under the architecture 1M-30 with helix antennas. In the case of the measurements with the modem

prototypes, the required values of transmission power were −73.49 dBm in the emulator and −84.74 dBm in the battery.

- The designed demonstrator (testbed and modem prototypes) for the wireless in-battery communication allows easy and fast further investigation of different methods for the physical and MAC layers. The models should be customized only in software.

9.3 Future Work

Although the requirements and targets of the IntLiIon project have been satisfactorily achieved in this work, there are some aspects that could be further investigated. Some recommendations for the future work are:

- The antennas employed in this work were either off-the-shelf evaluation boards or customized designs based on the off-the-shelf antennas. A more detailed study of antenna designs for this particular approach should optimize the system performance.
- The study of the influence of the system parameters on the channel could be extended. For example, bigger battery dimensions could be considered, e.g. in the order of magnitude of the big channel emulator. Also larger amounts of slave antennas.
- A more detailed modeling of the in-battery channel environment for the software CST MWS would allow a better matching between simulations and measurements in an automotive battery. More exact information about the internal geometry and dimensions of the battery would be helpful.
- Further types of messages (e.g. emergency notifications) should be implemented

Bibliography

[1] **Amoroso, F.** *Use of DS/SS signaling to mitigate Rayleigh fading in a dense scatterer environment*. In: *IEEE Personal Communications* 3.2 (1996), pp. 52–61.

[2] **Ashtiani, C.** *Analysis of Battery Safety and Hazards' Risk Mitigation*. In: *ECS Transactions* 11.19 (2008), pp. 1–11.

[3] *AWGN Channel, Matlab Dokumentation*. http://de.mathworks.com/help/comm/ug/awgn-channel.html, [Last access: June 2016]. MathWorks. 2015.

[4] **Barker, R.** *Group synchronizing of binary digital systems*. In: *Communication theory* (1953), pp. 273–287.

[5] **Bloessl, B.** *gr-foo Module*. https://github.com/bastibl/gr-foo, [Last access: August 2015]. 2015.

[6] **Bogusch, R. L.** *Digital communications in fading channels: Modulation and coding*. In: *Mission Research Corp., Santa Barbara, CA, Report no. MRC* (1989).

[7] **Borwein, P.** and **Mossinghoff, M.** *Barker sequences and flat polynomials*. In: *London Mathematical Society Lecture note series* 352 (2008), p. 71.

[8] **Brandl, M., Gall, H., Wenger, M., Lorentz, V., Giegerich, M., Baronti, F., Fantechi, G., Fanucci, L., Roncella, R., Saletti, R., Saponara, S., Thaler, A., Cifrain, M.**, and **Prochazka, W.** *Batteries and battery management systems for electric vehicles*. In: *Design, Automation Test in Europe Conference Exhibition (DATE), 2012*. 2012, pp. 971–976.

[9] **Carr, J.** and **Hippisley, G.** *Practical Antenna Handbook 5/e*. 5th. McGraw-Hill/TAB Electronics, 2011.

[10] **Computer Simulation Technology AG**, *CST Microwave Studio: Workflow and Solver Overview.* 2014.

[11] **Coonrod, J.** *Understanding When To Use FR-4 Or High Frequency Laminates*. In: *Onboard Technology* (2011), pp. 26–30.

[12] **Costas, J. P.** *Synchronous communications*. In: *Proceedings of the IRE* 44.12 (1956), pp. 1713–1718.

[13] **Couch II, L. W.** *Digital and Analog Communication Systems*. 6th. Upper Saddle River, NJ, USA: Prentice Hall PTR, 2000.

[14] **Cristi, R.** *Wireless Communications with Matlab and Simulink: IEEE802.16 (WiMax) Physical Layer*. Monterey, CA 93943, 2009.

[15] *Datasheet: Black Chip*. Laird Technologies.

[16] **Degardin, V., Lienard, M., Degauque, P., Simon, E.**, and **Laly, P.** *Impulsive noise characterization of in-vehicle power line*. In: *IEEE Trans. on Electromagnetic Compatibility* 50.4 (2008), pp. 861–868.

[17] **Dominicis, C. M. D., Ferrari, P., Flammini, A.**, and **Sisinni, E.** *Evaluation of synchronization methods by means of software defined radio architectures*. In: *International Symposium on Precision Clock Synchronization for Measurement, Control and Communication*. 2009, pp. 1–6.

[18] **Dostert, K.** *Vorlesungsskriptum: Stoerresistente Informationsuebertragung*. KIT. 2015.

[19] **Etacheri, V., Marom, R., Elazari, R., Salitra, G.**, and **Aurbach, D.** *Challenges in the development of advanced Li-ion batteries: a review*. In: *Energy & Environmental Science* 4.9 (2011), pp. 3243–3262.

[20] **Ettus Research**, *Application Note: Synchronization and MIMO Capability with USRP Devices*. Tech. rep.

[21] **European Broadcasting Union (EBU)**, *The use of band III in Europe*. 2014.

[22] **Frenzel, L.** *Fundamentals of Communications Access Technologies: FDMA, TDMA, CDMA, OFDMA, AND SDMA*. In: *Electronic Design* (2013).

[23] **Gardner, F.** *A BPSK/QPSK Timing-Error Detector for Sampled Receivers*. In: *IEEE Transactions on Communications* 34.5 (1986), pp. 423–429.

[24] **Gene Berdichevsky Kurt Kelty, J. S.** and **Toomre, E.** *The Tesla Roadster Battery System*. In: (2006).

[25] **George, M., Hamid, M.**, and **Miller, A.** *Gold code generators in Virtex devices*. In: *Xilinx Application Note xapp217, v1* 1 (2000).

[26] **Gibson, E. M.** *The Mobile Communications Handbook*. Ed. by **Gibson, J. D.** 2nd. Boca Raton, FL, USA: CRC Press, Inc., 1998.

[27] **Gnann, T.** and **Ploetz, P.** *Status Quo und Perspektiven der Elektromobilitaet in Deutschland*. Working paper sustainability and innovation. Fraunhofer ISI, 2011.

[28] *GNU Radio Wiki*. http://gnuradio.org/redmine/projects/gnuradio/wiki/, [Last access: May 2016]. GNU Radio. 2016.

[29] **Goldsmith, A.** *Wireless Communications*. Cambridge University Press, 2005.

[30] **Guimaraes, D. A.** *Digital transmission: a simulation-aided introduction with VisSim/Comm*. Springer Science & Business Media, 2010.

[31] **Harris, F. J., Dick, C.,** and **Rice, M.** *Digital receivers and transmitters using polyphase filter banks for wireless communications.* In: *IEEE Transactions on Microwave Theory and Techniques* 51.4 (2003), pp. 1395–1412.

[32] **Hiebel, M.** *Fundamentals of Vector Network Analysis.* Rohde & Schwarz, 2011.

[33] **Instruments, N.** *Pulse-Shape Filtering in Communications Systems.* In: (2014).

[34] **International Organization for Standardization (ISO),** *ISO 11898-6:2013. Road vehicles. Controller area network (CAN). Part 2: High-speed medium access unit.* 2013.

[35] **International Telecommunication Union (ITU),** *Report ITU-R SM.2180: Impact of industrial, scientific and medical (ISM) equipment on radiocommunication services.* Tech. rep. 2010.

[36] **Jondral, F.** *Nachrichtensysteme: Grundlagen, Verfahren, Anwendungen; mit 160 Bildern und 10 Tabellen.* 4., Ueberarb. und erweiterte Aufl. Wilburgstetten: Schlembach, 2011.

[37] **Khanam, F., Mitra, S., Hossain, M.,** and **Karmokar, D.** *Multiband low profile Modified Inverted-FL strip antenna for 5.2/5.8 GHz WLAN and 5.5 GHz WiMAX applications in laptop computer.* In: *15th International Conference on Computer and Information Technology (ICCIT).* 2012, pp. 598–603.

[38] **Kraus, J.** *Antennas.* McGraw-Hill, 1988.

[39] **Krutak, R., Raimund, W., Jellinek, R., Zopf-Renner, C., Emmerling, B., Figenbaum, E., Hjorthol, R., Bendsen, H., Marbjerg, G.,** and **Skov, R. S. H.** *State of the Art Electric Propulsion: Vehicles and Energy Supply.* Tech. rep. Compett, 2013.

[40] **Lee, M., Lee, J., Lee, I., Lee, J.,** and **Chon, A.** *Wireless battery management system.* In: *World Electric Vehicle Symposium and Exhibition.* 2013, pp. 1–5.

[41] **Lee, W. Y.** *Mobile cellular communications.* 1989.

[42] **Liflaender, J.** *Ceramic Chip Antennas vs. PCB Trace Antennas: A Comparison.* In: *Microwave Product Digest* (2010).

[43] **Linhart, R.** and **Fiala, P.** *Symbol Synchronization for SDR Using a Polyphase Filterbank Based on an FPGA.* In: *Radioengineering* (2015).

[44] **Lint, J. H. van.** *Introduction to coding theory.* Second. Berlin: Springer-Verlag, 1992.

[45] **Liu, L., Cheung, S., Yuk, T.,** and **Weng, Y.** *Cable Effects on Measuring Small Planar UWB Monopole Antennas.* INTECH Open Access Publisher, 2012.

[46] **Loh, T. H.** and **Alexander, M.** *New facility for minimally invasive measurements of electrically small antennas*. In: *Antennas and Propagation Conference, 2008. LAPC 2008. Loughborough*. 2008, pp. 313–316.

[47] **Loy, M.** and **Sylla, I.** *ISM-Band and Short Range Device Antennas*. Tech. rep. Texas Instruments, 2005.

[48] **Lu, L.**, **Han, X.**, **Li, J.**, **Hua, J.**, and **Ouyang, M.** *A review on the key issues for lithium-ion battery management in electric vehicles*. In: *Journal of Power Sources* 226 (2013), pp. 272–288.

[49] **Lu, X.**, **Liu, Y.**, and **Kim, H.** *Analysis of Cable Effects in Portable Antennas*. In: *Progress In Electromagnetics Research Symposium Proceedings*. Kuala Lumpur, MALAYSIA, 2012, pp. 1483–1486.

[50] **Luise, M.** and **Reggiannini, R.** *Carrier frequency recovery in all-digital modems for burst-mode transmissions*. In: *Communications, IEEE Transactions on* 43 (1995), pp. 1169–1178.

[51] **Lyons, R.** *Understanding Digital Signal Processing*. Pearson Education, 2010.

[52] **Malsbury, J.** *Modular, Open-source Software Transceiver for PHY/MAC Research*. In: *Proceedings of the Second Workshop on Software Radio Implementation Forum*. SRIF '13. Hong Kong, China: ACM, 2013.

[53] **Malsbury, J.** *Message Passing API & PHY/MAC with USRP and GNU Radio*. In: Ettus Research. 2013.

[54] **Markoff, J.** *Pursuing a Battery So Electric Vehicles Can Go the Extra Miles*. The New York Times. 2009.

[55] **Mathew, S.**, **Prakash, R.**, and **John, P.** *A smart wireless battery monitoring system for electric vehicles*. In: *2012 12th International Conference on Intelligent Systems Design and Applications (ISDA)*, 2012, pp. 189–193.

[56] **Mazar, H.** *An Analysis of Regulatory Frameworks for Wireless Communications, Societal Concerns and Risk: The Case of Radio Frequency (RF) Allocation and Licensing*. Universal Publishers, 2009.

[57] **Mengali, U.** and **D'Andrea, A. N.** *Synchronization techniques for digital receivers*. Applications of communications theory. New York, London: Plenum Press, 1997.

[58] **Miao, G.**, **Zander, J.**, **Sung, K.**, and **Slimane, S.** *Fundamentals of Mobile Data Networks*. Fundamentals of Mobile Data Networks. Cambridge University Press, 2016.

[59] **Mitola, J.** *The software radio architecture*. In: *IEEE Communications Magazine* 33.5 (1995), pp. 26–38.

[60] **Mohindra, R.** *Direct conversion receiver*. US Patent 5,438,692. 1995.

[61] **Molisch, A.** *Wireless Communications*. Wiley - IEEE. Wiley, 2010.

[62] **Mueller, K.** and **Muller, M.** *Timing Recovery in Digital Synchronous Data Receivers*. In: *IEEE Transactions on Communications* 24.5 (1976), pp. 516–531.

[63] **Nesimoglu, T.** *A review of Software Defined Radio enabling technologies*. In: *2010 10th Mediterranean Microwave Symposium*. 2010, pp. 87–90.

[64] **Oerder, M.** and **Meyr, H.** *Digital filter and square timing recovery*. In: *IEEE Transactions on Communications* 36.5 (1988), pp. 605–612.

[65] **Ouannes, I.**, **Nickel, P.**, and **Dostert, K.** *Cell-wise monitoring of Lithium-ion batteries for automotive traction applications by using power line communication: battery modeling and channel characterization*. In: *Proceedings of the 18th IEEE International Symposium on Power Line Communications and its Applications*. Glasgow, Scotland, 2014, pp. 24–29.

[66] **Ouannes, I.**, **Nickel, P.**, **Opalko, O.**, and **Dostert, K.** *Monitoring of Lithium-Ion Batteries in Electric/Hybrid Vehicles using Power Line Communication*. In: *Proceedings of the 6. GMM-Fachtagung Automotive meets Electronics*. Dortmund, Germany, 2015.

[67] **Ouannes, I.**, **Nickel, P.**, **Bernius, J.**, and **Dostert, K.** *Physical Layer Performance Analysis of Power Line Communication (PLC) applied for Cell-Wise Monitoring of Automotive Lithium-Ion Batteries*. In: *Proceedings of the 18th International OFDM Workshop*. 2014, pp. 1–8.

[68] **Pandeya, N.** and **Temple, N.** *AN-492: Selecting an RF Daughterboard*. Ettus Research. 2016.

[69] **Parkinson, B.** and **Spilker, J.** *Global Positioning System: Theory and Applications*. Progress in astronautics and aeronautics Bd. 1. American Institute of Aeronautics & Astronautics, 1996.

[70] **Pless, V.** and **Huffman, W. C.** *Fundamentals of error-correcting codes*. In: (2003).

[71] **Pozar, D.** *Microwave engineering*. Wiley, 1997.

[72] **Proakis, J.** *Digital Communications*. McGraw-Hill Series in Electrical and Computer Engineering. Computer Engineering. McGraw-Hill, 2001.

[73] **R. Garcia-Valle, J. A. P. L.**, ed. *Electric Vehicle Integration into Modern Power Networks*. Springer New York, 2013.

[74] **Rappaport, T.** *Wireless Communications: Principles and Practice*. 2nd. Upper Saddle River, NJ, USA: Prentice Hall PTR, 2001.

[75] **Research, E.** *USRP N200/N210 Networked series Datasheet*. 1043 North Shoreline Blvd, Suite 100, Mountain View, CA 94043, 2012.

[76] **Rice, M.** *Digital Communications: A Discrete-time Approach*. Pearson Prentice Hall, 2008.

[77] **Robert Bosch GmbH,** *Data and power - two become one*. In: *Press releases: Mobility Solutions* (2013).

[78] **Rondeau, T.** *GNU Radio Blog*. http://gnuradio.squarespace.com/, [Last access: August 2015]. Rondeau Research. 2014.

[79] **Roscher, V., Schneider, M., Durdaut, P., Sassano, N., Pereguda, S., Mense, E.,** and **Riemschneider, K.-R.** *Synchronisation using wireless trigger broadcast for impedance spectroscopy of battery cells*. In: *2015 IEEE Sensors Appl. Symposium (SAS)*. 2015, pp. 1–6.

[80] **Schneider, M., Ilgin, S., Jegenhorst, N., Kube, R., Puttjer, S., Riemschneider, K.,** and **Vollmer, J.** *Automotive battery monitoring by wireless cell sensors*. In: *2012 IEEE International Instrumentation and Measurement Technology Conference (I2MTC)*. 2012, pp. 816–820.

[81] **Shell, C., Henderson, J., Verra, H.,** and **Dyer, J.** *Implementation of a wireless battery management system (WBMS)*. In: *IEEE International Instrumentation and Measurement Technology Conference (I2MTC)*. 2015, pp. 1954–1959.

[82] **Sklar, B.** *Rayleigh fading channels in mobile digital communication systems I: Characterization*. In: *IEEE Commun. Mag.* 35.7 (1997), pp. 90–100.

[83] **Sklar, B.** *Rayleigh fading channels in mobile digital communication systems .II. Mitigation*. In: *IEEE Communications Magazine* 35.7 (1997), pp. 102–109.

[84] **Steiger, W.** *Future Mobility. Volkswagen Group Solutions for Sustainable Mobility*. 2014.

[85] **T. Schunder F. Freund, J. W.** *Robuste, skalierbare Bustopologien fuer LiIon-Batteriemanagementsysteme*. In: *U.R.S.I. Landesausschuss in der Bundesrepublik Deutschland Kleinheubacher Tagung*. Miltenberg, Germany, 2015.

[86] **Technologies, L.** *Datasheet:ANT-433-HETH*.

[87] **Turin, G.** *An introduction to matched filters*. In: *IRE Transactions on Information Theory* 6.3 (1960), pp. 311–329.

[88] *Tutorial: Programming Topics*. http://gnuradio.org/redmine/projects/gnuradio/wiki/Guided_Tutorial_Programming_Topics, [Last access: August 2015]. GNU Radio. 2015.

[89] *USRP Hardware Driver and USRP Manual*. http://files.ettus.com/manual/, [Last access: May 2016]. Ettus Research. 2016.

[90] **Van Bladel, J.** *Electromagnetic Fields*. IEEE Press Series on Electromagnetic Wave Theory. Wiley, 2007.

[91] **Wade, P.** *Understanding Circular Waveguide Experimentally*. In: *QEX* (2001).

[92] **Wallace, R.** *Design Note DN031: CC-Antenna-DK and Antenna Measurements Summary*. Tech. rep. Texas Instruments, 2010.

[93] **Wallace, R.** *Application Note AN058: Antenna Selection Quick Guide*. Tech. rep. Texas Instruments, 2013.

[94] **Weng, Y.**, **Cheung, S.**, and **Yuk, T.** *Effects of ground-plane size on planar UWB monopole antenna*. In: *TENCON 2010 - 2010 IEEE Region 10 Conference*. 2010, pp. 422–425.

[95] **Wong, K.** *Planar antennas for wireless communications*. Wiley series in microwave and optical engineering. Hoboken: Wiley, 2003.

[96] **Wood, J.** *A Glimpse of the Future*. In: *IEEE Microwave Magazine* 13.7 (2012), pp. 60–69.

[97] **Wu, Y.**, **Liao, X.**, **Chen, W.**, and **Chen, D.** *A Battery Management System for electric vehicle based on Zigbee and CAN*. In: *2011 4th International Congress on Image and Signal Processing (CISP)*. Vol. 5. 2011, pp. 2517–2521.

[98] **Zayed, N. M.** *Design and Simulation of Communications and Radar Systems*. In: ed. by **Mathworks**, 2014.

[99] **Zheng, M.**, **Qi, B.**, and **Wu, H.** *A Li-ion battery management system based on CAN-bus for electric vehicle*. In: *3rd IEEE Conference on Industrial Electronics and Applications (ICIEA)*. 2008, pp. 1180–1184.

[100] **Zimmer, G.** *Highly Accurate Hybrid Electric Battery Monitor*. In: *Power Electronics Europe* Issue 8 (2012), pp. 25–28.

[101] **Zoltowski, M.** *Equations for the Raised Cosine and Square-Root Raised Cosine Shapes*.

List of Publications

[102] **Alonso, D.**, **Opalko, O.**, and **Dostert, K.** *Channel Measurements and Simulations with Planar Inverted F-Antennas in an Enhanced Testbed for a Wireless Battery Management System*. In: *Proceedings of the 19th International ITG Workshop on Smart Antennas*. Ilmenau, Germany, 2015, pp. 1–8.

[103] **Alonso, D., Gavilan, J.**, and **Dostert, K.** *Software Defined Channel Sounder for Power Line Communications*. In: *KIT Scientific Working Papers*. Vol. 44. KIT, Karlsruhe, 2016.

[104] **Alonso, D., Opalko, O.**, and **Dostert, K.** *Physical Layer Performance Analysis of a Wireless Data Transmission Approach for Automotive Lithium-Ion Batteries*. In: *Proceedings of the 2015 IEEE Vehicular Networking Conference*. Kyoto, Japan, 2015.

[105] **Alonso, D., Opalko, O.**, and **Dostert, K.** *Parametrization of Automotive Lithium-Ion Batteries and Its Influence on the Wireless In-Battery Channel*. In: *10th European Conference on Antennas and Propagation (EuCAP)*. Davos, Switzerland, 2016.

[106] **Alonso, D., Winkler, C., Opalko, O.**, and **Dostert, K.** *Prototyping of the Physical and MAC Layers of a Wireless Battery Management System*. In: *Proceedings of the 2015 IEEE Vehicular Networking Conference*. Kyoto, Japan, 2015.

[107] **Alonso, D., Opalko, O., Sigle, M.**, and **Dostert, K.** *Towards a wireless battery management system: evaluation of antennas and radio channel measurements inside a battery emulator*. In: *Proceedings of the 80th IEEE Vehicular Technical Conference*. Vancouver, Canada, 2014, pp. 1–5.

[108] **Opalko, O., Alonso, D.**, and **Dostert, K.** *A 3-D EM Simulation Model of Rogowski Coils as PLC Coupling Elements for Traction Batteries*. In: *Proceedings of the 9th Workshop on Power Line Communications*. Klagenfurt, Austria, 2015.

[109] **Opalko, O., Alonso, D.**, and **Dostert, K.** *Measurements on Rogowski coils as coupling elements for power line communication in traction lithium-ion batteries*. In: *Proceedings of the IEEE 2015 International Symposium on Power Line Communications and its Applications*. Austin, Texas, USA, 2015, pp. 29–34.

[110] **Opalko, O., Alonso, D.**, and **Dostert, K.** *Modeling of the Power-line Channel in Automotive Li-ion Batteries with Rogowski Coils as Coupling Elements*. In: *Proceedings of the IEEE 2016 International Symposium on Power Line Communications and its Applications*. Bottrop, Germany, 2016.

[111] **Opalko, O., Simon, B., Alonso, D.**, and **Dostert, K.** *Physical Layer and Multicarrier Analysis for Power Line Communication Networks in Li-ion Batteries for Electric and Hybrid Vehicles*. In: *Proceedings of the 2015 IEEE Vehicular Networking Conference*. Kyoto, Japan, 2015, pp. 243–250.

List of Supervised Theses

[112] **Bernhardt, S.** *Software-Defined-Radio-basierte Implementierung von Codierungsverfahren fuer ein drahtloses In-Batterie-Kommunikationssystem*. Industrial Information Technology (IIIT), Karlsruhe Institute of Technology (KIT), 2016.

[113] **Gaitan Gutierrez, F.** *Simulation of an OFDM System for a Wireless Battery Management System*. Industrial Information Technology (IIIT), Karlsruhe Institute of Technology (KIT), 2014.

[114] **Gavilan Garcia, J. R.** *Implementation of Channel Sounder Techniques by means of Software Defined Radio*. Industrial Information Technology (IIIT), Karlsruhe Institute of Technology (KIT), 2014.

[115] **Kleis, R.** *Design eines RF-Front-Ends zur Funkkommunikation*. Industrial Information Technology (IIIT), Karlsruhe Institute of Technology (KIT), 2014.

[116] **Lax Cortina, J.** *FSK transceiver implementation by means of reprogramming the FPGA of the USRP*. Industrial Information Technology (IIIT), Karlsruhe Institute of Technology (KIT), 2014.

[117] **Loch, M.** *Software Defined OFDM System for wireless In-Batter y Communication*. Industrial Information Technology (IIIT), Karlsruhe Institute of Technology (KIT), 2016.

[118] **Wabo Kogaing, A.** *Simulation und Analyse des Kanals fuer Li-Ionen Batterien in Elektrofahrzeugen*. Industrial Information Technology (IIIT), Karlsruhe Institute of Technology (KIT), 2015.

[119] **Winkler, C.** *Software-Defined-Radio-basierte TDMA Implementierung fuer drahtlose In-Batterie-Kommunikation*. Industrial Information Technology (IIIT), Karlsruhe Institute of Technology (KIT), 2015.

Forschungsberichte aus der Industriellen Informationstechnik (ISSN 2190-6629)

Institut für Industrielle Informationstechnik
Karlsruher Institut für Technologie (KIT)

Hrsg.: Prof. Dr.-Ing. Fernando Puente León, Prof. Dr.-Ing. habil. Klaus Dostert

Band 1 Pérez Grassi, Ana
Variable illumination and invariant features for detecting and classifying varnish defects. (2010)
ISBN 978-3-86644-537-6

Band 2 Christ, Konrad
Kalibrierung von Magnet-Injektoren für Benzin-Direkteinspritzsysteme mittels Körperschall. (2011)
ISBN 978-3-86644-718-9

Band 3 Sandmair, Andreas
Konzepte zur Trennung von Sprachsignalen in unterbestimmten Szenarien. (2011)
ISBN 978-3-86644-744-8

Band 4 Bauer, Michael
Vergleich von Mehrträger-Übertragungsverfahren und Entwurfskriterien für neuartige Powerline-Kommunikations-systeme zur Realisierung von Smart Grids. (2012)
ISBN 978-3-86644-779-0

Band 5 Kruse, Marco
Mehrobjekt-Zustandsschätzung mit verteilten Sensorträgern am Beispiel der Umfeldwahrnehmung im Straßenverkehr. (2013)
ISBN 978-3-86644-982-4

Band 6 Dudeck, Sven
Kamerabasierte In-situ-Überwachung gepulster Laserschweißprozesse. (2013)
ISBN 978-3-7315-0019-3

Band 7 Liu, Wenqing
Emulation of Narrowband Powerline Data Transmission Channels and Evaluation of PLC Systems. (2013)
ISBN 978-3-7315-0071-1

Die Bände sind unter www.ksp.kit.edu als PDF frei verfügbar oder als Druckausgabe bestellbar.

Band 8 Otto, Carola
Fusion of Data from Heterogeneous Sensors with Distributed Fields of View and Situation Evaluation for Advanced Driver Assistance Systems. (2013)
ISBN 978-3-7315-0073-5

Band 9 Wang, Limeng
Image Analysis and Evaluation of Cylinder Bore Surfaces in Micrographs. (2014)
ISBN 978-3-7315-0239-5

Band 10 Michelsburg, Matthias
Materialklassifikation in optischen Inspektionssystemen mithilfe hyperspektraler Daten. (2014)
ISBN 978-3-7315-0273-9

Band 11 Pallauf, Johannes
Objektsensitive Verfolgung und Klassifikation von Fußgängern mit verteilten Multi-Sensor-Trägern. (2016)
ISBN 978-3-7315-0529-7

Band 12 Sigle, Martin
Robuste Schmalband-Powerline-Kommunikation für Niederspannungsverteilernetze. (2016)
ISBN 978-3-7315-0539-6

Band 13 Opalko, Oliver
Powerline-Kommunikation für Batteriemanagement-Systeme in Elektro- und Hybridfahrzeugen. (2017)
ISBN 978-3-7315-0647-8

Band 14 Han, Bin
Characterization and Emulation of Low-Voltage Power Line Channels for Narrowband and Broadband Communication. (2017)
ISBN 978-3-7315-0654-6

Band 15 Alonso, Damián Ezequiel
Wireless Data Transmission for the Battery Management System of Electric and Hybrid Vehicles. (2017)
ISBN 978-3-7315-0670-6